T0257756

Sustainable Development: Studies in Urban Development and Tourism

Sustainable Development: Studies in Urban Development and Tourism

Edited by **Kane Harlow**

New York

Published by Callisto Reference,
106 Park Avenue, Suite 200,
New York, NY 10016, USA
www.callistoreference.com

Sustainable Development: Studies in Urban Development and Tourism
Edited by Kane Harlow

© 2015 Callisto Reference

International Standard Book Number: 978-1-63239-581-8 (Hardback)

This book contains information obtained from authentic and highly regarded sources. Copyright for all individual chapters remain with the respective authors as indicated. A wide variety of references are listed. Permission and sources are indicated; for detailed attributions, please refer to the permissions page. Reasonable efforts have been made to publish reliable data and information, but the authors, editors and publisher cannot assume any responsibility for the validity of all materials or the consequences of their use.

The publisher's policy is to use permanent paper from mills that operate a sustainable forestry policy. Furthermore, the publisher ensures that the text paper and cover boards used have met acceptable environmental accreditation standards.

Trademark Notice: Registered trademark of products or corporate names are used only for explanation and identification without intent to infringe.

Printed in the United States of America.

Contents

Permissions

List of Contributors

Preface

This book is a result of research of several months to collate the most relevant data in the field.

The technological improvement of our civilization has established a consumer society growing more rapidly than the planet's resources allow, with the needs for resource and energy rising exponentially in the past century. For the assurance of a secured future for humans, an enhanced comprehension of the environment as well as technological solutions, behaviors and mindsets is important, in accordance with modes of development that the ecosphere of earth can support. Sustainable development provides an approach that would be useful to blend with the managerial strategies and assessment tools for decision and policy makers at the regional planning level. Thus, the book covers important chapters grouped under two sections, "Policy and Sustainable Urban Development" and "Sustainable Tourism".

When I was approached with the idea of this book and the proposal to edit it, I was overwhelmed. It gave me an opportunity to reach out to all those who share a common interest with me in this field. I had 3 main parameters for editing this text:

1. Accuracy – The data and information provided in this book should be up-to-date and valuable to the readers.

2. Structure – The data must be presented in a structured format for easy understanding and better grasping of the readers.

3. Universal Approach – This book not only targets students but also experts and innovators in the field, thus my aim was to present topics which are of use to all.

Thus, it took me a couple of months to finish the editing of this book.

I would like to make a special mention of my publisher who considered me worthy of this opportunity and also supported me throughout the editing process. I would also like to thank the editing team at the back-end who extended their help whenever required.

Editor

Part 1

Policy and Sustainable Urban Development

Sustainable System Modelling for Urban Development Using Distributed Agencies

Bogart Yail Marquez, Ivan Espinoza-Hernandez
and Jose Sergio Magdaleno-Palencia
COLEF and ITT
México

1. Introduction

Developing a sustainable simulation system consists essentially of generating sustainable, artificial worlds with the capacity to produce results similar to those observed in the real world. This allows for varying parameters in a controlled, reusable experimental environment, something that cannot be easily achieved through mathematical models. The field of simulation is broad and multidisciplinary and has had an impressive growth since the 90's. While the area of simulation has been expanding to new horizons in traditional systems research, there are yet a series of unsolved epistemological issues David et al. (2010).

On the other hand, the social sciences face challenges that go beyond their capabilities of processing information. By using modern techniques such as computer agents and other methodologies, it is possible to aid in the testing and the formulation of theories Davidsson (2002).

Computer agent techniques are having a greater acceptance in recent years in different fields of science, and as a result, they have begun to be implemented as a simulation technique. Agent techniques consist on using small, independent programs called agents that are modeled to represent the social actors, be it people, organizations or corporations. Agents are designed to react to changes in their environment, which is also modeled to represent real world conditions that the actors would encounter in the given situation of interest Gilbert (2007).

A fundamental characteristic of agent based models is the ability for agents to interact, that is, they are able to transmit informative messages to other agents and can act based on the information received. Messages can represent spoken dialog between people or other indirect forms of communication. Information on actions such as observation of other agents or the perception of actions taken by other agents can also be acquired through messages. When modeling computer agents, specifying how they handle their interactions with other agents and the environment is one of the main differences with other computational models Gilbert (2007).

The complexity level of using these techniques increases as the number of agents increases. Even though it has been mentioned in the multi-agent community the need to develop and implement methodologies, surprisingly, very little has been done and therefore many areas of science have been excluded.

The motivation for this work is the need to establish a methodology for the study of sustainable systems in situations where conventional analysis cannot provide satisfactory information on the complexities of social phenomena and social actors. In general, the proposed methodology describes the use of several computational techniques and interdisciplinary theories. This growing consensus must be capable of describing every aspect of a sustainable system, as well as serve as a common language in which different theories can be juxtaposed.

1.1 Sustainable systems

Sustainability refers to the equilibrium between a species and the resources in its environment. By extension, this can apply to the exploitation of resources bellow the renovation limit.

Sustainability is generally associated with the definition of sustainable development, which refers to being able to satisfy the needs of the present without compromising the ability to satisfy the needs of the future generations. The concept of sustainability applies to the systems composed of human beings and nature. The structures and functionality of the human component in terms of society, economy, and rights among others should be such that they self-enforce and promote the persistence of the structures and functionality of the natural components—such as the ecosystem, biodiversity and biogeochemical cycles—and vice versa Cabezas et al. (2005). Therefore, one of the research challenges on sustainability resides in the link between the form of functioning of the ecosystems towards the structures and the functionality of the associated social system. This is why the information theory based indicators can grasp the human nature and the elements of the system and make sense of the disparity of the variables in the system Márquez, Castañon Puga & Suarez (2010).

2. Sustainable system modeling

Sustainable development is about assuring a good quality of life for the present and future generations. This can be achieved through the three strands of social equality—which are social, economical, and environmental—which recognize each other's needs, can maintain stability in these levels—with special attention to economic development and employment—and responsibly manage the natural resources available while protecting the environment Márquez, Castanon-Puga, Castro & Suarez (2010). Sustainability is even necessary among systems to ensure coexistence. As an example, the economic performance in regard to the expense of the community is not sustainable; without effective environmental protection, the economic activities will be obstructed. Sustainability does not require a perfect solution; it is in essence a goal or a vision that organizations should strive to achieve Ciria (2009).

Studies have been done on this focus such as the ones on sustainable agriculture, which is a philosophy that guides the development of agriculture systems in a multidisciplinary way in the areas of economy, environment, and social impact. Sustainable agriculture requires a global focus, one that is oriented towards solving the problems of the food industry and fibers industry Williams & Dollisso (1998).

And so, liking the different levels that are required in order to create a sustainable system is a challenge that is yet to be solved, from a countries economy in relation to its available resources and the existing population to the availability of those resources to individuals and their economic status. These social, economical and environmental variables are analyzed with a bottom-up approach based on how a social structure functions.

2.1 Social system

When dealing with social systems, certain basic characteristics in the organization must be met. One of these is that the consequences of the social systems are probabilistic and non-deterministic. Furthermore, as human behavior is not entirely predictable due to its complexity, dealing with consumers, suppliers, regulation agencies and others cannot wait for a predictable behavior Suarez et al. (2007). Organizations are seen as systems within systems. Said systems are complex, producing a whole that cannot be understood by only analyzing the individual parts. They must be dealt with as a system that is characterized by all the essential properties of any social system Yolles (2006). For this reason, the following properties must be taken into consideration when modeling organizational systems:

2.1.1 Interdependent pieces

A change in any of the components will have an effect on the other components. The external and internal interactions of the system reflect different stages of control and autonomy.

2.1.2 Homeostasis or firm state

An organization can achieve a firm state only when two requirements are met, unidirectionality and progress. Unidirectionality means that in spite of changes, the same results or established conditions will be obtained. Progress referring to the desired outcome, is a degree of progress that is within the set boundaries determined as tolerable.

2.1.3 Borders or limits

It is the marker that determines what is inside and what is outside the system. It need not be a physical marker; it consists of a closed area surrounding the selected variables that have the most interaction with the system.

2.1.4 Morphogenesis

The organizational system, distinct from the mechanical and biological system, has the capacity to modify its basic structure. This ability is identified by Buckley as its main identifying characteristic Boulding (1956).

One of the objectives of this research is to predict social behavior by using models. Social behavior is a behavior that favors those that conform to the group, producing cooperation and self-organization Jaffe & Zaballa (2010). According to Ross Ashby Ashby (2004), the word organization has a multitude of meanings, specifically, its use in the areas of computation and neural science is of great importance to this research. In social systems, the question arises of what is the behavior of individuals when in a group (cities, groups or networks), and why they exhibit such behavior.

2.2 Economic systems

Another component in a sustainable system is the economy that reigns in a city, the market economy. Salary rates are normally regulated by contracts and are subjected to the market's rules in the middle and long term. Goods and services needed for daily urban life are also affected by the market's rules.

The industrialization process and the concentration of investment due to work specialization and the use of economies of scale generate the process of urbanization. The activities that thrive in the urban center generate job positions that are primarily occupied by the locals but also attract outsiders that are looking for better conditions. This generates a cycle that leads to sustained population growth and the demand of public services, which in turn requires taxation to keep up with demand, improved service and proper administration of tax revenue.

2.3 Environmental systems

As it is now, the market economy does not always lead to an efficient allocation of resources in the provision of public services. In order to determine optimal distribution of public investment, it is necessary to have a cost-benefit analysis, prioritizing the social aspects and considering the externalities, tending towards a balance between economies of agglomeration and diseconomies produces by clustering. To exemplify, the investments in basic sanitation (potable water and sewage systems) should not be weighted on the basis of the end-user's income, but instead on the benefit produced by lowering the mortality and disease rates which increases productivity in the population, income and quality of life.

3. Distributed agency modeling

Agent based models are an increasingly potent tool in social systems simulations as they can represent phenomena that is difficult to describe using other mathematical formalisms. However, these models have had a limited involvement in formulating social systems owing to the fact that their distinct abilities are more useful in situations where the future is unpredictable. In said situations, traditional analysis methods applied to simulation models are less efficient in the decision making process. The use of models such as policy simulators provides significant aid in taking decision in the public and private sector. This is of special relevance as these models have had to date limited impact in influencing decisions.

The application of agent based models in studying heterogeneous behavior has been successful as it allows for each agent to have different information, different rules and be faced with different situations that allows the study of the behavior at a macro level in the global system.

This modeling technique has been used to combine the anthropological data on the behavior of individuals and groups in society with detailed information of the effect of climate change on the environment Lempert (2002).

When faced with a complex sustainability problem, such as deciding what actions need to be taken to deal with global climate change, a broad range of possible scenarios must be considered. At the least, a rigorous analysis needs a way to identify and define the most important and likely scenarios. Advanced made in viable agent model simulations has also allowed new methods in decision analysis to adapt to these types of problems. Uncertainty arises when parts of a decision will not or cannot agree over one or several key components in a decision analysis to be used in non predictable models such as: the system model, the a priori statistics of any parameter describing the system model, and the value of the function used to classify the model's results. These multi-scenario simulation models provide a systematic and quantitative orientation for which scenarios information should be reviewed and extracted.

Although the use of agents in the social sciences has been stated in the field of artificial intelligence Gilbert (2007), as it is one of the first areas to have studied this topic Russell & Norvig (2004) and more precisely, in distributed systems. By themselves, agents are not enough to model a real social system, nevertheless, distributed agency based systems is an active area of research with promising results in the fields of engineering and social sciences. These types of systems also reduce the barrier between physical and sociological systems as the perceived view of the world is nonlinear.

It must be stated that this research does not use the conventionally defined agent model—which defines agents as atomic concepts or actors—but instead uses the distributed agent model or distribution agents—which does not define independent actors but instead considers the organism that extends throughout the whole of the system. Agents can be any process, it can change any system based on the independently contained information.

The idea behind a distributed organism modeling language derives from a vision of the world in which appearance is omnipresent, where compounds are irreducible to their components and exist in different dimensions where different rules apply Suarez et al. (2008). Distributed agencies attempt to solve problems between groups of agents, finding the solution within the result of the cooperative interaction between agents.

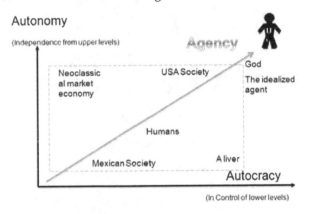

Fig. 1. Levels agents represented hierarchically

Communication facilitates cooperation; the degree of cooperation between agents can be from full cooperation to hostile. In the first case there is a high cost for the total communication between agents. In the second case some agents could block the objectives of others. To have the cooperation and coordination mechanisms in an agent system succeed, an additional system must exist that enables the members of the system to reach agreements when each individual agent is defending its own interests. This system should reach mutually beneficial solutions, taking into consideration all points of view. Such a system is known as negotiation Gilbert (2007).

Applications of this technology are considered very useful for distributed industrial systems development such as process control, e.g., automatic management of intelligent buildings with private security and resource management. Other areas have developed applications for air traffic control used in airports like Sydney, Australia Julian & Botti (2000). Distribution agents is a promising strategy that can correct an undesirable centralized architecture Russell & Norvig (2004). Throughout the focus of traditional multi-agent systems and utility maximization, actors choose the best alternative given the set of possibilities that is found in each level.

The main distinction from the proposed focus is that the phase space includes the transformations made by an upper level. On the other hand, an agent is composed of subcomponents belonging to a lower level that can possess their own agencies. It is an agent's responsibility to present its subcomponent's individual phase spaces with optimal solutions that are acceptable to the parent upper level agents. In other words, agents found in subcomponents optimize the phase spaces in their parent agents, while the parent agents must consider the manipulation of this world of possibilities in order to reach the desired global behavior. To this effect, if an agent were to be considered a corporation, this level would be composed of the subdivisions that form the company, and these in turn are directed by groups of people. The company as a whole is also located in a level that is ruled by legislation relating to industrial practices, which are a component of an upper level that forms a specific society.

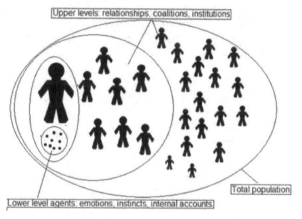

Fig. 2. Multiple levels of identity of distributed agency

4. Case study

Urban simulation that considers sustainability has remained an interesting topic in research for many years. Aspects such as urban growth, congestion, and segregation have a high demand in advanced modeling focuses. Each of the focuses and techniques that have been applied present advantages and drawbacks. Even as aggregation techniques have been criticized for their poor results in these types of models, they have been receiving renewed attention in recent times Benenson & Torrens (2004). Among these techniques, the agents based models are considered the most promising as they provide a detailed understanding of the structure and processes of urban systems Márquez, Manuel & Saurez (2010). Combining these techniques with geographic information systems (GIS) will greatly improve urban simulation.

There has been greater acceptance of agent modeling of urban systems in the last decades Gilbert (2007). Agent based models and artificial societies are very similar being the same techniques in dynamic systems, cellular automatons, genetic algorithms, and distributed agent systems. The differences are centered in the simulation of systems and in research program design Drennan (2005).

The location being studies is Ciudad Juarez, a Mexican city in the northern part of the state of Chihuahua in a region known as El Paso del Norte and bordering with El Paso, Texas in the United States of America. Its geographic environment can be delimited by the municipality of Juarez, which extends for 3,599 km^2.

The city is settled between the Sierra de Juarez and the valley of Juarez in a geographical area historically formed by fluvial deposits originating from the stream of the Rio Bravo. Its terrain is rugged to the west over the hills of the Sierra de Juarez and with smooth slopes with an east to west direction in the valley area. The heights of the most elevated terrains located in the Sierra de Juarez are above 1,800 meters over the mean sea level (msl). The inhabited area over the hills in the mountainous range consists of elevations between 1,250 and 1,350 msl. Most of the urban sprawl is located between the elevation of 1,150 and 1,200 msl and distributed in the valley of Juarez and extending to the south.

Therefore, for our proposed work-in-progress case study, if we consider a municipality an agent, this upper-level agent is composed by subcomponents, which in our case study of the city of Juarez, Mexico, will be represented by the AGEBS that compose this city. AGEBS is the terminology used to describe the different areas of the city that are in turn are composed of neighborhoods. The data set of the city of Juarez is divided into 549 areas, known as AGEBS. "The urban AGEB encompass a part or the totality of a comunity with a population of 2500 inhabitants or moreâĂe in sets that generally are distributed in 25 to 50 blocks" INEGI (2006)

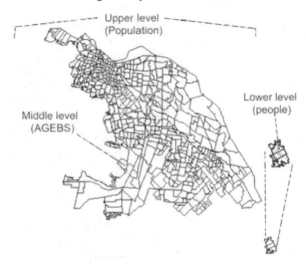

Fig. 3. Levels of agents represented on the City of Juarez

A city has several qualities that align with the definition of complexity. That is why performing a simulation of a city requires the study of a complex system and emergence. The use of simulations in the study of a city's urban growth helps perform social experiments while avoiding costs and risks. Simulation tools are already readily available that apply different techniques and models to study growth. This research applies the distributed agency methodology on a sustainable system for a city located in Mexico, creating a model of a sustainable system for urban growth as a secondary objective.

To develop the simulation methodology, the concept of a city and its influencing processes must be understood. The concept of a city according to Camagni Camagni (2004) is based on generalizing process that begins from the historical and geographical existence of the cities and continues to consider the city as a significant whole, an autonomous socioeconomic entity Camagni (2004). A city constitutes a production entity, in which a group of goods and services are internally produced; all of these internal and external processes that engulf it can be represented by distributed agencies offering the ability to represent the surrounding environment, take autonomous actions and simulate actions such as consumption and productive activities among others. The use of distributed agencies to create an urban simulation describes satisfactorily the processes of cooperation, communication, and decisions.

5. Proposed methodology

The methodology to be implemented represents an innovative focus on creating a simulation architecture. Named Distributed Agency (DA), this methodology represents a general theory of collective behavior and the formation of structures; it redefines the level of agency in two forms. Primarily, there are no obvious agents; each of these entities that represent an emergent force is the result of organized sub-agents in the lower levels. In the second form, agents can belong to different levels . The language of distributed agency expresses the observed behavior as the result of agents maximizing their objective functions Suarez et al. (2009).

This research intends to develop a sustainable system methodology using various mathematical and computational theories that are not conventionally used in the social sciences and provide a new focus for the creation of computer simulation architectures. The research shows how the DA methodology in combination with other techniques can be used to simulate social behavior, using agents with limited reasoning capacity and complex interactions. The simulations expand the knowledge available on social complexity, setting the basis for a nonlinear methodology to study the scenarios that have been developed using existing traditional methods. With this multi-focus study, it is intended to show how agents interact in their environment, their behavior and the relationships between different levels imitating the ones found in the real world.Márquez, Castanon-Puga, Castro & Suarez (2011). The DA methodology consists of eight steps that are :

5.1 Determining the agency levels and their relations

This phase analyses the existing relations in the social system and determines the levels of the system. In order to accomplish this, the problems that need to be solved are identified and their functions are described for each level in a physical frame. The decision and parameter input and output variables are also identified. An intrinsically holistic philosophy must be pursued without reducing the system to its basic components, since no phenomena can exist by itself in a sustainable system, where each node is defined by its link with other nodes Heylighen (2008). That is why it is necessary to establish the objective functions of each level of agency and the prevalent nodes and links.

Taking the case study of the sustainable development of Ciudad Juarez, the total population, and all related factors such as immigration and birth rate, is to be analyzed at a macro level with a top–down approach. The micro level interactions such as spouse selection, decision to start a family, and number of children depending on the education and social level is analyzed with a bottom–up approach.

To achieve this reproduction modeling a macro system, dynamic systems are used. A dynamic system allow the representation of all the elements and relations of the sustainable system's structure and the evolution of the system in time Márquez, Castañon Puga & Suarez (2010). It also outputs the mathematical equations of the macro level model, the results of these equations define the characteristics of the macro agency level. Another part will be determining the mid level agencies, an agent's actions in its environment and the relation between agencies this way allowing the observation of micro level cases.

Some proposed classifications that have been defined by researchers involved in urbanism define a structure with different layers or levels, depending on the interactions and the different structures. The graph shown in figure 1 has been referred to as "Camagni's wedding cake" which shows three layers or levels (international, regional, and local) and various structures (hierarchical, non hierarchical, and mixed). The elements in each layer are interrelated, forming a network in each level, in a similar way cities are interrelated forming a complex set of link.

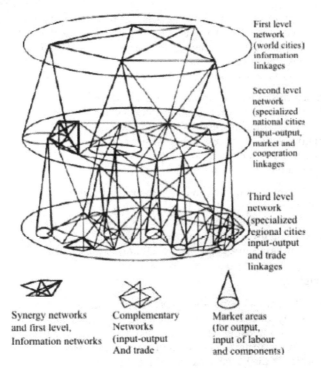

First level network (world cities) information linkages

Second level network (specialized national cities input-output, market and cooperation linkages

Third level network (specialized regional cities input-output and trade linkages

Synergy networks and first level, Information networks

Complementary Networks (input-output And trade

Market areas (for output, input of labour and components)

Fig. 4. Camagni's wedding cake

The highest level in this case study is the sustainable system, the proceeding three levels are the economic, environmental and social systems.

5.1.1 Social systems

A city's urban growth is composed of "vegetative growth + migration balance". Both elements are studied by the tools provided by demographics. A city's growth is driven by economic

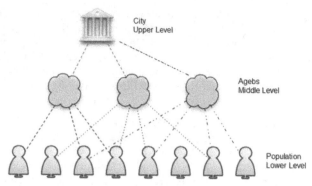

Fig. 5. Multiple levels represented

growth. Growth and optimal size of a city can be studied through simplified theoretical models. With the growth of population, the scale of production and job market rises, and the technological development and public service efficiency increase. Simultaneously, the diseconomy also increases, leading to higher unemployment, congestion, pollution, crime and social distortion. These factors are detrimental externalities that have little effect on the ones taking the decision Márquez, Castanon-Puga, Magdaleno-Palencia & Suarez (2011).

With the appearance of industrialization, the development process and population growth in cities is accelerated. This is a job creating process but also demands services since the people that fill the job positions should be located close to their place of work. These people will demand housing, urban services, food, clothing and furniture among other goods and services. Urban agglomerations then arise with the demand of the inhabitants to perform their activities and receive goods and services.

Aside from vegetative growth, the urban phenomenon is bolstered by migration flows. These flows are made up of people that are constantly arriving in cities looking for better conditions and opportunities. They generally have a rural background or originate from less developed countries. Because of the broad effect that population has in the development of a city, the total population of Ciudad Juarez is the variable that is extracted from the social system.

5.1.2 Economic systems

Economic theory states that as population increases, the scale of production and job market increases. In the study of a city's urban growth, it is important to analyze the process from its foundation without losing sight of important events as the industrialization process, migration flows. Factors such as rent rates, public service demand are intrinsically intertwined with job creation which is the variable extracted for the economic system.

5.1.3 Environmental systems

The variable that is taken into account for this system is the water supply being an essential part of any economy and society. Therefore, the sustainable management of this resource is a necessary condition for a sustainable society and economy. The sustainable use of water is defined as the use of an amount capable of sustaining a society and can develop in an indefinite future without altering the integrity of the hydrological system or the ecosystems that depend on it Gleik et al. (1996) .It is increasingly difficult to achieve this balance, but still, to approximate sustainable growth, all converging factors must be studied.

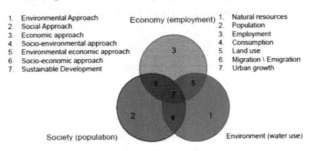

Fig. 6. Sustainable system

5.2 Data mining

The continuous increase of available information, originating from existing projects such as data bases necessary for the simulation of sustainable systems makes the use of data mining indispensible. Sustainable development requires a great deal of data to generate reliable models. To determine the data relevant to the three inherent systems of sustainability, it is necessary to review real statistical and geographical data originating from government institutions such as the National Institute of Statistic and Geography (INEGI by its name in Spanish, Instituto Nacional de Estadística y Geografía) and the National Population Council (CONAPO, by its name in Spanish, Consejo Nacional de Población). These institutions provide the necessary quantitative information for the social system. Data for the economic system is provided by the National Survey on Occupation and Employment (ENOE by its name in Spanish, Encuesta Nacional de Ocupación y Empleo) and for the environmental system the data is obtained from the Municipal Committee of Water and Disinfection of Juarez (JMAS by its name in Spanish, Junta Municipal de Agua y Saneamiento de Juarez).

Data mining is an implicit method of extracting information, such as weather patterns, with the intention of gaining knowledge Dubey et al. (2004). Significant progress has been achieved in this field during the last fifteen years; most of the research effort has been focused on the development of efficient algorithms capable of extracting knowledge from data, leaving the philosophical basis neglected Peng et al. (2008). The selection and processing of information leads to the use of high performance computing, with exploits such as social simulation and tools that give meaning and use to the information obtained. For this reason, it is important to pay attention to the conceptual frames and use it as the basis for developing the proposed methodology.

5.3 Rule generation

Using the Neuro-Fuzzy system to automatically generate the necessary rules, this data extraction phase using a fuzzy system becomes complicated as it is necessary to determine the necessary rules and what variables to consider. Implementing the Nelder-Mead (NM) search method, being more efficient than other methods such as genetic algorithms, more precise and compact models can be created as it was demonstrated in other experiments ?. It is a numerical method designed to minimize an objective function in a multi-dimensional space, approximately searching for an optimal local solution in an N variable problem when the objective function has smooth variations Stefanescu (2007).

To generate the rules, the following markers must be considered:

5.3.1 Total population

Population growth, as previously mentioned, consists of "vegetative growth + migration balance" illustrated in the following formula:

$$PT = [N - M + I - E] \tag{1}$$

Where:

PT: Total population

N: Birth rate

M: Mortality rate

I: Immigration

E: Emigration

5.3.2 Employment

To measure employment, the result of both the work force and total population is considered. Employment rate is determined by fifteen variables, one for each AGEB .

$$FL = \frac{PEA}{PT} \tag{2}$$

$$Po = P2 + P3 + P4 + P5 + P6 + P7 + P8 + P9 + P10 + P11 + P12 + P13 + P14 \tag{3}$$

$$TE = \frac{Po}{PEA} \tag{4}$$

Where:

FL: Work Force

PEA: Economically Active Population

PT: Total Population

TE: Employment Rate

Po: Occupied Population

P2: Occupied population in the secondary sector

P3: Occupied population in the tertiary sector

P4: Occupied population as employee or working-class

P5: Occupied population as day laborers

P6: Occupied population that is self-employed

P7: Occupied population that works up to 32 hours a week average

P8: Occupied population that works from 33 to 40 hours a week average

P9: Occupied population that works 41 to 48 hours a week average

P10: Occupied population that does not receive compensation for their work

P11: Occupied population that works or less than the monthly minimum wage

P12: Occupied population with an income of 1 to 2 minimum monthly salaries

P13: Occupied population with an income of 2 to 5 minimum monthly salaries

P14: Occupied population with an income greater than 5 minimum monthly salaries

PEA: Economically Active Population

5.3.3 Water consumption

Considering the proposed markers for water consumption made in studies by Cervera Cervera (2007) and proposing environmental damage variables based on JMAS, INEGI, the XII general population census and Vivienda 2000 , the following equation is obtained :

$$D = VD_1 + VD_2 \tag{5}$$

$$CA = \frac{VAFUDM}{TOVP} - D \tag{6}$$

Where:

VAFUDM/year = Annual volume of water billed for domestic use in cubic meters = 115,633,582.

VAFUDL/day = Annual volume of water billed for domestic use in liters per day.

TOVP = Total number of occupants in residence.

D: Environmental damage or degradation

VD1: Private residences with pluming connected to sewage, ravine, river, lake or sea.

VD2: Private residences without water, pluming or electricity.

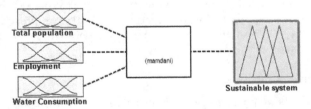

Fig. 7. Sustainable system

Most models with agents applied to natural resource management are structured with two elements, the agents that represent the entities in the modeled system and a simple cellular automaton as the spatial representation. The sole use of cellular automatons in general has limited the modeling possibilities since this abstraction process can be restrictive Galán-Ordax et al. (2006). By combining different modeling techniques, more realistic representations can be obtained, which is why the initiative to integrate fuzzy logic to extract rules from statistical data in data bases, all is needed is to input the equation and the necessary agency rules will be generated.

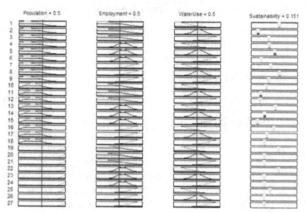

Fig. 8. Generated sustainability rules

5.4 Distributed agency model

Existing relations are very important in complex systems modeling, as they intertwine the system. A phenomenon can only be conceived in relation to another phenomenon and no phenomenon can exist by itself. The nodes are defined by their relations with other nodes and links through which they connect. This is an intrinsically holistic philosophy, it is not possible to reduce a system to individual components Heylighen (2008).

Undertaking the simulation of a sustainable system implies a holistic analysis, carrying out a multi-level analysis. The goal is to establish a mechanism in which different levels can be referenced within a reality with a general methodology. Each level is different from the rest, this means that by grouping several agents from a lower level, this group will behave as a single entity.

The implemented methodology represents a new approach to creating a simulation architecture. This distributed Agency (DA) methodology represents a general theory of collective behavior and the formation of structures. The DA approach treats agents as something agent-like, contrasting with traditional approaches where entities are or are not considered agents Suarez et al. (2009) .

5.5 Implementation

Months of work can be required to gather information, build, verify, and validate models, to design experiments, evaluate, and interpret results. The cost of a simulation is high, as it depends on the gathering of different types of information, from qualitative to quantitative. The initial foundation work and maintenance of simulation capabilities involves having trained personnel, software, and hardware among other costs Benenson & Torrens (2004). Another issue faced is the use of a tool that could simulate the different levels of a model in a single software.

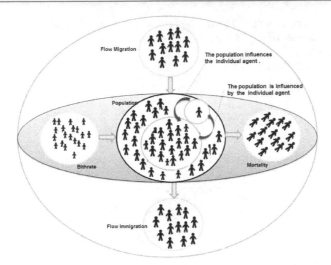

Fig. 9. Bottom-up and top-down model of the population

As an example, in the simulation of Ciudad Juarez, to represent the lowest levels of agency, 1,313,338 independent particles (the city's population in 2005) and their interactions must be managed if the upper level is to be used endogenously. The model that is presented in this work is based on the main sustainable relations between demographics, employment, the consumption of potable water, and the changes in land use caused by these factors.

Different aspects must be considered in order to choose a suitable platform. Among them is an orientation to creating agent based models which is necessary to simulate continuous events; most platforms are event based. The platform must also be configurable in various aspects such as the having individual selection and job management. Lastly, it must ease the development process, allowing researchers to quickly test models, theories and strategies in areas with dynamic and complex simulations.

Using the NetLogo platform, it is possible to simulate social phenomena, model complex systems and give instructions to hundreds or millions of independent agents all acting holistically Wilensky (1999). It also permits the use of a geographical information system with special and statistical data. These features make it possible to explore the relation and behavior of agents and the emergent patterns that arise from the interactions within a geographical space. NetLogo can be defined as a programming language for the modeling of multi-agent systems integrated with a social and natural phenomena simulation. The NetLogo environment can simplify exploring emergent phenomena Vidal (2007), and is also suitable for the modeling of complex systems varying in time, allowing or independent instructions to be given to the agents at the same time. The mentioned aspects can give the opportunity to discover the link between the micro level behavior of the individuals and the macro level patterns that arise from the interactions of the individuals Wilensky (1999).

5.6 Model validation

Real world simulations that include the population as an objective must include some form of validation. In econometrics, there is abundant data to verify population and economic studies while in other areas such as anthropology, there is a shortage of data. The supply of this data is a secondary concern. The main concern is for the data sets to adapt to an agent's architecture. An example of this is the study that centered on the cognitive origins from social theory Drennan (2005).

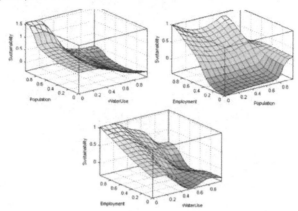

Fig. 10. Graphic representing the selected variables to measure the sustainable system

Aside from the population growth caused by vegetative growth and migration flows, employment is a marker that better reflects the relation between the economic and demographic factors.   The primary occupation of Ciudad Juarez comes from the manufacturing sector (maquiladoras), which drives the changes in land use. By creating the maquiladoras, the demand for industrial usage has increased. With an increased job offer, migration increases and thus the population increases Romo et al. (2009). In consequence, the demand for residential land use is increased and results in the creation of new residential districts.

The use of natural resources such as water depends on the unique properties of the city. Ciudad Juarez has a dry arid climate, an annual mean temperature of $17.3\,^{o}$ C , an annual mean precipitation of 223.8mm, an average of 48.1 rainy days and an average of 1.8 snow days Sánchez (1997).

To evaluate the model of Ciudad Juarez, it is a fast growing city with an inherent demand for land and public services that has overwhelmed the urban planning schemes, in particular, in environmental issues such as water consumption. This growth phenomenon puts stress on the environment, manifesting itself as an increase in waist production, excessive gas emissions, vehicle congestion, and other effects. It all contributes to the degradation of the environment with effects in the air, ground and water Romo et al. (2009).

The superficial waters of the Rio Bravo that enter Ciudad Juarez are used in their entirety for irrigation in the valley of Juarez, an annual supply of 60 thousand acre-feet or approximately 74 million mm^3. The Rio Grande is the only renewable source of water for the Ciudad Juarez-El Paso region; Ciudad Juarez is entirely dependent on the aquifer called Hueco Bolson. Water extraction as increased in recent years, with an annual average rate of 2.5%. In 1990, 119.8 mm^3 of water was extracted, in the year 2000, an extraction of 153 mm^3 was reported and by the year 2005 the rate was 147.3 mm^3 . The approximate annual extraction is 175 mm^3 , this pumping provides a service capacity of 330 liters per inhabitant per day.

It is estimated that the Hueco Bolson aquifer has a surface of 260.89 acres (the approximate extension of the urban sprawl of Ciudad Juarez) JMAS (1997), and has an annual recharge rage of 35 mm^3 . This means that the extraction rate in Ciudad Juarez is approximately five times greater than the recharge rate provided by rainfall JMAS (2005). Nevertheless, the aquifer also receives subterranean recharges in a north-south hydraulic gradient coming from the U.S. side. These recharges have not been properly quantified.

5.7 Simulation and optimization

The initial simulation process has been carried out in two tiers, i.e., the macro and the micro levels. To illustrate we use the macro model of the dynamic systems from the top-down model. The dynamic systems allow us to depict all the elements and relationships from the sustainable system's structure. In addition, we will be able to visualize missing links or connections between the entities, and therefore adjust the corresponding mathematical equations in the model.

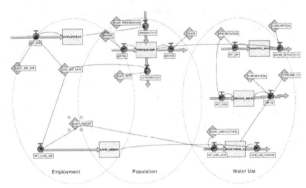

Fig. 11. Macro level sustainability implemented in NetLogo

On the other hand, obtaining the most relevant data will permit the representation of higher-level agents. In the context of this study, these will be: social, economic and environmental agents striving for harmony in all the aspects. The modeling process of this project is based upon methods and mathematical expressions, such that represent the theoretic behavior of the land use in the proposed representation. The simulation is presented in a geographical space using the NETLOGO framework. This software proves to be useful in representing geographical systems as well. Using agents during the modeling process provides a better comprehension of the structure and processes in urban systems. The integration of modeling with geographical information systems has dramatically improved the possibilities of urban simulations. We can visualize agents at the meso level , in the way they are affected by their surroundings and relationships within the same levels and the adjacent levels, i.e., superior and inferior levels.

5.8 Output data analysis

In the first part of the presented urban simulation model, the focus is on the social-environmental aspect. It can be observed how the sharp demographic growth from Ciudad Juarez has incremented along with the consumption of drinking water. Ciudad

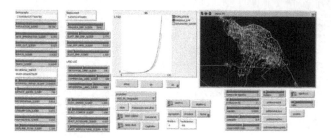

Fig. 12. Sustainability's meso level, implemented in NetLogo

Juarez's water system obtains its entire water supply from the Hueco Bolson aquifer, with an extraction rate that exceeds by many times the natural replenishment rate. Based on this information, the simulation model determined that in twenty years' time this resource will be insufficient to provide city's water needs. In parallel, the urban model simulation found that Ciudad Juarez is strongly linked to the American economy due to its nature of border town, situation that may lead in the future to financial crises. However, there is still data and relationships not clearly defined and required in the model. If the existing variables are accounted for in each system and linked between each other, an interesting output is obtained which need to be examined in the respective disciplines (social, economic and environmental). It should be noted that developing a sustainable model using different simulation techniques could prove to be a valuable instrument during planning stages. The usage of dynamic systems supported the construction of new theoretical models and contributed to expand our understanding of the connections within the systems. This partly to the availability of a global perspective of the situation and some concrete instances provided by the relationships in the lower agency levels.

6. Conclusion

Developing a methodology for a sustainable system may imply the use of different techniques, theories and the researcher's perceptions which often differ between each other. Usage of a unique statistical methodology in this context could be insufficient given the requirements of encompassing a complex system. The proposed methodology is developed using a holistic approach, analyzing the diverse levels of dimension and time, applying nonlinear dynamics where required, harnessing the emergent properties of the model, and the self-organizing processes and interactions between several levels in the manifold respective dimensions. The interdisciplinary nature of this field sets the main goal when using this methodology, to explore the relationship between the diverse social and computational theories linked to complexity sciences. The outcome of the use of the methodology provides a reliable alternative to complement, substitute and expand the traditional approaches in the context of social sciences from the point of view in which complex systems are studied to developing the techniques proposed in this work. Multidisciplinary connections and multi-leveled modeling is still an unexplored field in computational social sciences and in the context of social simulations as well. Currently, this methodology has proven to be a way to achieve further advances in the objectives set by scholars in the area of social studies. The accuracy of statistics used in social sciences, can be improved by extending the number of variables, and its validity is kept when analyzed at a single level, however, most of the social, economic and environmental issues are part of a higher complex system. Thus, it is a difficult task to embrace a general methodology for any complex system, not only by reason of the multiplicity of variables suitable for measurement, but also due to the nonlinear dynamics, self-organization and interactions between levels and dimensions. Hence, analyzing complex systems under the approach of complex systems and multi levels is greatly required. The use cases presented here can be linked together to the interactions between multiple levels only if the most significant relationships are clearly identified. The methodology applied to a sustainable system may imply a vast amount of information with different theories and computational techniques. The presented sustainable system used a diversity of simulation techniques being these key instruments for planning efforts of any type. The use of dynamic systems helped create new theoretic models and understand the underlying relationships within the system, visualizing these outcomes globally. Distributed agencies were helpful to represent particular use cases and the interpersonal associations between agencies. The proposed methodology was developed holistically, the analysis of the sustainable system, was studied at a macroscopic level determining all the processes in between job opportunities, population and water sources. This analysis provides a model that is simultaneously dependent and influential of lower levels. The intermediate hierarchies are considered given that most of the analysis ranges between boundaries without accounting for the middle sections. This can be accomplished by keeping in mind the relationships between components at different levels.

7. References

Ashby, R. (2004). Principles of the self-organizing system, *E:CO Special Double Issue* 6: 102–126.

Benenson, I. & Torrens, P. (2004). *Geosimulation, Automata-based Modeling of Urban Phenomeno*, Geosimulation, John Wiley and Sons, London.

Boulding, K. (1956). General systems theory the skeleton of science, *Management Science* 6(3): 127–139.

Cabezas, H., Pawlowski, C., Mayer, A. & Hoagland, T. (2005). Sustainable systems theory: ecological and other aspects, *Journal of Cleaner Production* 13(5).

Camagni, R. (2004). Urban economics, *Investigaciones regionales*, N. 5 pp. 235–237.

Cervera, L. (2007). Indicadores de uso sustentable del agua en Ciudad Juárez, Chihuahua, *Red de Revistas Científicas de América Latina y el Caribe, España y Portugal* .

Ciria (2009). Sustainable construction procurement.

David, N., Caldas, J. C. & Coelho, H. (2010). Epistemological perspectives on simulation iii, *Journal of Artificial Societies and Social Simulation* 13(1): 14.

Davidsson, P. (2002). Agent based social simulation: A computer science view, *Journal of Artificial Societies and Social Simulation* 5.

Drennan, M. (2005). The human science of simulation: a robust hermeneutics for artificial societies, *Journal of Artificial Societies and Social Simulation* 8(1).

Dubey, P., Chen, Z. & Shi, Y. (2004). Using branch-grafted r-trees for spatial data mining.

Galán-Ordax, J. M., López-Paredes, A. & del Olmo-Martínez, R. (2006). Modelado y simulación basada en agentes con sig para la gestión de agua en espacios metropolitanos.

Gilbert, N. (2007). *Agent-Based Models*, Sage Publications Inc., Los Angeles.

Gleik, P. H., Postel, S. L. & Morrison, J. I. (1996). The sustainable use of water in the lower colorado river basin.

Heylighen, F. (2008). Five questions on complexity.

INEGI (2006). Ii conteo de población y vivienda 2005. instituto nacional de estadística geografía e informática.

Jaffe, K. & Zaballa, L. (2010). Co-operative punishment cements social cohesion, *Journal of Artificial Societies and Social Simulation* 13: 4.

JMAS (1997). Proyecto de las plantas de tratamiento de aguas residuales norte y sur y obras complementarias de alcantarillado de cd. juárez, chihuahua.

JMAS (2005). Plan de trabajo 2005, gobierno del estado de chihuahua.

Julian, V. & Botti, V. (2000). Agentes inteligentes: el siguiente paso en la inteligencia artificial, *Horizonte 2025* .

Lempert, R. (2002). Agent-based modeling as organizational and public policy simulators.

Márquez, B. Y., Castañon Puga, M. & Suarez, E. D. (2010). On the simulation of a sustainable system using modeling dynamic systems and distributed agencies, *in* N. C. (INC) (ed.), *INC2010: 6th International Conference on Networked Computing*, IEEE, pp. 86–90.

Márquez, B. Y., Castanon-Puga, M., Castro, J. R. & Suarez, D. (2011). Methodology for the modeling of complex social system using neuro-fuzzy and distributed agencies, *Journal of Selected Areas in Software Engineering (JSSE)* .

Márquez, B. Y., Castanon-Puga, M., Castro, J. R. & Suarez, E. D. (2010). On the modeling of a sustainable system for urban development simulation using data mining and distributed agencies.

Márquez, B. Y., Castanon-Puga, M., Magdaleno-Palencia, J. S. & Suarez, E. D. (2011). Modeling the employment using distributed agencies and data mining.

Márquez, B. Y., Manuel, C. n.-P. & Saurez, D. (2010). Sustainable system simulation for urban development using distributed agencies.

Peng, Y., Kou, G., Shi, Y. & Chen, Z. (2008). A descriptive framework for the field of data mining and knowledge discovery, *International Journal of Information Technology and Decision Making* vol 7: 639–682.

Romo, L., Córdova, G., Brugués, A., Rubio, R., Ochoa, L., Díaz, I., Lizárraga, G., Guerra, E., Aniles, M., Sapién, A., Mota, M. & Fong, J. (2009). Zonificación y ordenamiento ecológico y territorial del municipio de juárez.

Russell, S. & Norvig, P. (2004). *Inteligencia Artificial. Un Enfoque Moderno*, 2 edn, Pearson Prentice Hall.

Sánchez, J. (1997). Monografía de ciudad juárez.

Stefanescu, S. (2007). Applying nelder mead's optimization algorithm for multiple global minima, *Romanian Journal of Economic Forecasting*. pp. 97–103.

Suarez, E. D., Castañón Puga, M. & Marquez, B. Y. (2009). Analyzing the mexican microfinance industry using multi-level multi-agent systems.

Suarez, E. D., Rodríguez-Díaz, A. & Castañón Puga, M. (2007). *Fuzzy Agents*, Vol. 154 of *Studies in Computational Intelligence*, Springer, Berlin, pp. 269–293.

Suarez, E. D., Rodriguez-Diaz, A. & Castanon-Puga, M. (2008). *Soft Computing for Hybrid Intelligent Systems*, Vol. 154, Springer, chapter Fuzzy Agents, pp. 269–293.

Vidal, J. M. (2007). *Fundamentals of Multiagent Systems with NetLogo Examples*.

Wilensky, U. (1999). Netlogo software.

Williams, D. & Dollisso, A. (1998). Rationale for research on including sustainable agriculture in the high school agricultural education curriculum, *Journal of Agricultural Education* 39, No. 3.

Yolles, M. (2006). Organizations as complex systems: An introduction to knowledge cybernetics.

2

Sustainable Urban Development Through the Empowering of Local Communities

Radu Radoslav, Marius Stelian Găman, Tudor Morar,
Ştefana Bădescu and Ana-Maria Branea
Faculty of Architecture,"Politechnica" University of Timişoara, Timişoara,
Romania

1. Introduction

The financial, economical, social and ecological crysis that violently outburst worlwide after 2008 is the result of structural challanges, such as globalization, climate changes, the pressure on resources, migrations, social exclusions, demographical changes, the ageing of population, mobility, etc., which all have a strong urban dimension, which was determined, at an international level, mainly by the "SPRAWL"-type growth (Saunders, 2005), which only encourages the economical side of the development. In order to overcome this crisis, we propose, as short, medium and long-term strategies, the analysis and the solutions that we found for the problems of the city of Timişoara (Romania). These studies take into consideration the works of C. Butters (Butters, 2004), who states that regional sustainable development cannot be achieved, and therefore neither that of each city, community or neighbourhood, without gradually improving all of the following aspects, at the same time: the social one, which brings social diversity, accessibility, identity, security, variety, involvement and sociability; the economical one, which can be achieved by cutting revenue expenditure, improving functions, diversifying activities and adjacent financial structures, services and communications, by management and flexibility; the ecologic one, through a more harmonious use of land, through biodiversity and bio-climate, by producing non-pollutant energy, re-naturalising the water cycle, recycling, adequate accessibility and by improved overall health.

The desire to exit this crysis has determined, throughout the majority of the European cities, new types of strategies, with objectives, development directions and clear measures, that are adopted, with small differences, all across the European Union, and that accentuate the importance of good governance at a European/national/regional/local level. From our point of view, however, good governance applied only to those four levels, as stated in the European documents, is not enough; we believe that the local level should be divided in a number of subunits. For the four initial levels of governance, the research methods are clearly formulated, according to the 2007 Leipzig Charta. Solving the problems regarding social exclusions, structural changes, the ageing of population, climate changes and mobility is the main theme of this European document, which hopes to lead to economic prosperity, social ballance and a healthy environment. In this document, the prosperity mentioned above depends on an increased attention paid to the subunits of the local level, which includes "*the underpriviliged neighbourhoods, in the context of a city as a whole*" (Leipzig Charter, 2007). It is clear that this act reffers to an integrative aspect of governance, which implies a harmonious relation between the inhabitant and its physical environment. The relations between the spatial units and the corresponding social units (Lang, 1994) form a Behaviour Setting. Between these Behaviour Settings, there is a continuous competition for occupying the best position in order to solve the economical and social differences, which depend on a certain type of governance. This competition has always been tempered by cooperation relationships, which are, most of the times, not planned. The European act that completes the Leipzig Charta, namely the 2010 Toledo Declaration, supports "*a good governance, based on the principles of transparency, of participation, of responsibility, of efficiency, of subsidiarity and of coherence*" (Toledo Declaration, 2010). The proposed hierarchy, namely the European/national/regional/local hierarchy, cannot stop here because good governance should reach the level of a group of inhabitants that live in appartaments served by the same staircase of a condominium building, passing through the Behaviour Settings levels of a District, a Neighbourhood (a Territorial Unit of Reference according to the Romanian legislation), a Vicinity Unit and of a Group of Housing Units (Radoslav et al., 2010a). This implies the implementation of two contemporary administrative principles, namely subsidiarity and procesuality. The first principle refers to establishing a connection between the decision and the level upon this decision has the most important effects, while the second aspect takes into consideration the open character of the options and of the decisions regarding territorial planning. Good governance,"*a more efficient and effective use of public resources*", should be provided in order to "*increase the direct public participation of the citizens*" (Toledo Declaration, 2010). The current Romanian legislation states that 30% of the taxes cashed in by the States' Budget should remain at the disposal of the central government, 26% should go to each county's government, while 44% should go to the local governments. When Romania will be reorganised from an administrative point of view, this distribution will most likely be modified: 10% of the taxes will remain at the disposal of the central government, 20% will go to the regional governments (a new administrative form), 30% to each county's government, while 40% will go to the local governments, which represents a step forward towards descentralization. The Toledo Declaration also supports "*an implication, a taking on tasks and a responsabilization of the factors, at multiple levels and from an integrative point of view*". What we propose is that 40-45% of the money that remain at the local level be redistributed to the subunits previously mentioned through the similar appliance of the principles of subsidiarity and procesuality.

All these are attractive generic sentences, but problems appear when we try to apply this statement in everyday life. *Where, who* and *how* can this desiderate be applyed? *Where* means the delimitation of an area, with a certain autonomy, *who* means the delimitation of a community with a certain identity (Radoslav, 2000) that operates within that area, while *how* refers to the governance of the area and of the community that form a Behaviour Setting thus delimitated. Therefore, we can speak of the Spatial Unit named Earth, that should harmonize, within a Behaviour Setting, with the Social Unit of earthlings, Europe with Europeans, Romania with Romanians, Banat (a region in Romania) with the people who live in it, Timis (a county in Romania) with those who inhabite it, Timisoara with its citizens. It is obvious that these delimitations are the result of a continuous historical process, in which the whole procedes the parts and in which the identity is being born where only homogeneity existed before (Alexander, 1987) and that these Behaviour Settings are made out of different parts, that function in a complex manner. In the competition between Behaviour Settings of the same level good governance plays a key role.

2. The relationship between larger Behaviour Settings (European, National, Regional and Local) and good governance

Our studies, developed within the Research Group for a Sustainable Territorial Development – "Politechnica" University of Timișoara, have concentrated on different levels of Behaviour Settings: European, Euroregional, Regional, County, Growth Pole, Municipality, District, Neighborhood, Vicinity Unit, Block or Group of Housing Units, for which we propose measures for good governance, according to the principles of the two European documents. In this chapter, parts of our studies will be presented, that justify the holistic triade – economic, social and ecological – as starting point for the transparency of information needed by all actors that operate in the area, especially by citizens, their participation in the subunits of the local level being absolutely neccessary.

2.1 The relationship between the Behavior Setting of the European Union and good governance

According to the first ESPON scenarios (when Romania was not part of the European Union), the city of Timișoara was situated outside the area of European integration with strong potential in Central Europe, which ended just after Budapest, and also outside the area of European integration with future potential, which started in Athens, passed through Sofia and ended at Bucharest. After Romania became part of the European Union, Timișoara received a very important part, according to the ESPON Cohesion based scenario for 2030 (ESPON 3.2, 2006), as hinge between the extensions of the two areas mentioned above (fig. 1). From a demographical point of view, many East-European cities have experienced decreases in population of approx. 15% in the last 20 years, percentage very close to that of Timișoara. By 2065, up to a third of the European population will be older then 65. Due to reduced fertility, high life expectancy and migration, the European Union will maintain its total population until 2050. But the structure of the population will change, because the number of youngsters and working people will decrease. According to the same ESPON study, our area is part of the regions with slow urbanization and decreased population. These phenomena, together with the mixing of population through enmigration and immigration, will lead to the possibility of loosing the identity, flexibility and diversity of an area. The decrease in population will lead to loss of usage efficiency of all types of existing infrastructure. The towns from the area can no longer ensure education, sports and health facilities, commerce, public transport, universities, etc. The implementation, through good governance, of the two principles (subsidiarity and procesuality) at all the levels of the Beahviour Settings, cannot be done without the support of the population, which should become an informed partner. Without good governance, the situation can aggravate in all sectors. A proposed measure of good governance is "polycentrism", which implies promoting some complementary and interdependant network of large cities, as well as medium and small ones that can lead to the integration of the rural environment, as alternatives to the metropolis or to the capital cities. In order to consciously accomplish this polycentrism, some common evaluation criteria of the towns (population, competitivity, connectivity, education system, innovation, etc.) are needed.

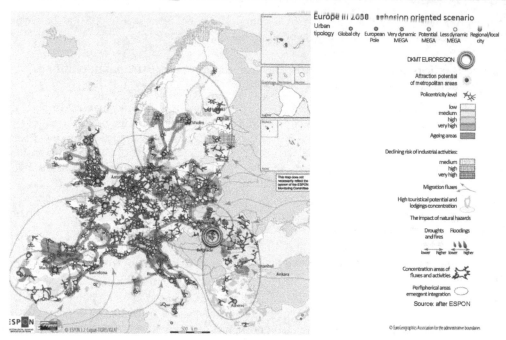

Fig. 1. The Behaviour Setting of Europe in 2030 – cohesion oriented scenario

2.2 The relationship between the Behavior Setting of the Danube-Cris-Mures-Tisa (DKMT) Euroregion and good governance

The next level of our studies refers to the good governance of the Behaviour Setting of the Danube-Cris-Mures-Tisa (DKMT) Euroregion (fig. 2), in which Timişoara plays the main role as a growth pole. The problems regarding the harmonization of the transfrontalier areas that compose this Behaviour Setting refer to the creation of equilibrium between the component regions of the European neighbourh countries. In the transfrontalier area Romania-Hungary-Serbia economical differences, as well as socio-political ones appear due to their development in the last 70 years, differences that have been accentuated by Romania's complete isolation after 1980 and by Sebia's after the 1990s. The organization of the European Union leads to the transformation of this transfrontalier area into a very important Behaviour Setting (fig. 3), whose purpose will be reached when equilibrium between the forces of the counties and regions of the three countries will be created. This is a very delicate issue, because if this situation is not fully understood, the desire to expand or to dominate the others will be very difficult to manage.

According to fig. 1, considering the attraction potential of the metropolitan areas, the Growth Pole Timişoara is closer to the atractivity of Belgrade. According to our study from 2007 (Radoslav et al., 2010b), the polycentric development of the DKMT Euroregion, composed at that time by two counties from Voivodina (Serbia), four counties from Hungary and four counties from Romania (later, the county of Hunedoara left this Euroregion), proposes a central superpole Timişoara-Arad, with more then 700.000 inhabitants, supported by Szeged and Novi Sad towards the West and by the inter-city

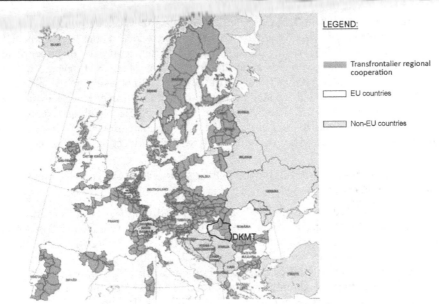

Fig. 2. The Behaviour Setting of the DKMT Euroregion and the other transfrontalier regions of cooperation in the Behaviour Setting of European Union

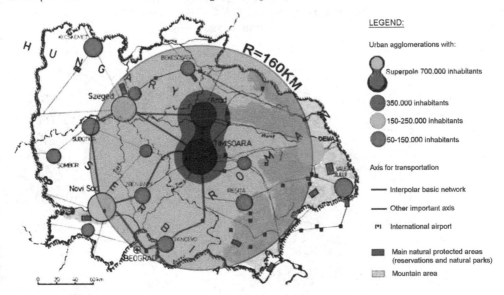

Fig. 3. The Behaviour Setting of the DKMT Euroregion

development area Deva-Hunedoara-Simeria (Radoslav et al., 2010b), named "Provincia Corvinia" towards East. The influence radius of this superpole is of approx. 5.000 km, which makes it a Euroregional pole. Each of the three supporting poles has a population between 170.000-250.000 inhabitants and gravitates at a distance of approx. 150 km from the superpole Timișoara-Arad. The Behaviour Setting of this superpole demonstrates that the influence radius of 30 km between the Growth Pole Timișoara and the Development Pole Arad overlap near the villages Vinga and Orțișoara, which compels their cooperation (Radoslav et al., 2010b) through the creation of a logistic pole – supporting pole at middle distance between Timișoara and Arad. This is the only way to strenghten and expand the transeuropean network – with special attention paid to the reduction of travels, expansion of general interest services in the rural and peripheral areas, ecological problems, as well as to the protection of farmland. Also, transfrontalier risk management can be promoted, including the impact of climate changes, by intensifying territorial cohesion politics.

2.3 The relationship between the Behavior Setting of the Western Development Region and good governance

The next level of the study refers to the harmonization through good governance of the Behaviour Setting of the Western Development Region, Romania (the former Banat region, between 1948-1964, currently a non-administrative development unit) (fig. 4). The Western Development Region has a surface of 32.034 sqkm (13,4% of the Romania's surface), and is composed by four counties (Arad, Caraș-Severin, Hunedoara and Timiș), with 42 towns (out of which 12 are cities) and 276 communes, with a total of 318 territorial-administrative units and a large number of villages abandobed in the last 20 years. The population of the region was 1.930.458 inhabitants in 2005 (with a decrease of 14% in the last 15 years) and a density of 61,1 inhabitants/sqkm. The urbanization percentage of the region is 63,6% urban population, larger than the national average of 54,9% (The Regional Development Agency for the Western Region, 2011). From an economical point of view, the region has an

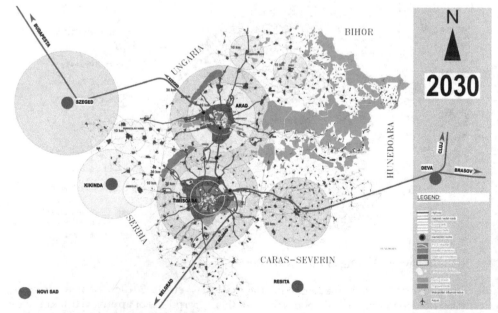

Fig. 4. The Behaviour Settings of the Growth Pole Timișoara and of the Development Pole Arad

estimated GDP for 2011 of approx. 1,3 billions euro (The Romanian National Prognosis Comitee, 2011), similar to most of the other eight Romanian development regions, but much less than regions with similar population from Europe. Considering the environment, the climate is temperate, continental and moderate. As a consequence of the global climate changes, between 1992-2002 a defficit of precipitation of 14,6 mm has been registred. The multiannual average temperature shows an increase of 0,5°C in the last 20 years. These phenomena have determined, in the last years, a series of storms and floodings, as well as warmer winters, with the resulting effects (The Regional Development Agency for the Western Region, 2011).

2.4 The relationship between the Behavior Setting of the Timiş County and good governance

The next level of the study refers to the harmonization through good governance of the Behaviour Setting of Timiş County. Considering the population, this Behaviour Setting had 658.837 inhabitants in 2005, with a decrease of 20.000 inhabitants since 2002, and a predicted decreasing trend until 2025. The structure of the population is even more interesting; 13,6% of the population is over 65, trend that will reach 17% in 2025. The average life expectancy is 71,43 years, while the active population represents 48%. This structure determines a very low unemployment rate, of only 2,3%, in comparison to the 7% national level (The Regional Development Agency for the Western Region, 2011). From an economical point of view, Timiş County is estimated to have the highest GDP of all Romanian counties in 2011, of approx. 600 millions euro, but also four times higher then the GDP of the poorest county from the Western Development Region – Caraş-Severin (The Romanian National Prognosis Comitee, 2011). From an ecological point of view, in Timiş County there has been an increase of the average temperature by 0,5°C in the last 20 years, that complicates even more the situation in the areas prone to natural hazards, especially in the areas with floods and landslides, unstable areas (situated towards East, North-East and South-East from Timişoara) and in the highly seismical area situated in the South of Timişoara (fig. 5). In the areas situated in the vicinity of villages where zootechnics was intensively developed, there

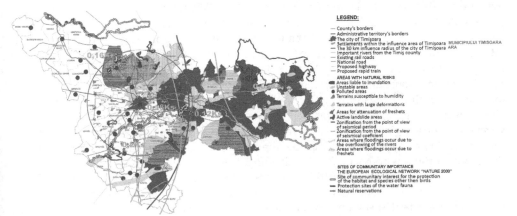

Fig. 5. The Behaviour Setting of Timiş County – areas with natural risks and sites of communitary importance

are some acute pollution problems, which imply an extension of the Growth Pole Timişoara with the exclusion of these areas. In Timiş County there also exist some protection areas. In our proposals, these areas should be amplified in order to mitigate part of the previously mentioned risks by planting a forrest belt (a green corridor).

2.5 The relationship between the Behavior Setting of the Growth Pole Timişoara and good governance

The next level of our study refers to the harmonization through good governance of the Behaviour Setting of the Growth Pole Timişoara, which is a juridical association of eight territorial administrative units around Timişoara. From the point of view of urbanization, it has been proposed a maximal growth of the built areas up to the traffic belt and afterwards only alongside the radial penetrations, in order to reduce land use and stop the uncontrolled expansion. The terrain yet unbuilt up to the traffic belt is proposed, in our studies, to be building prohibited, so that this areas become natural areas of protection. Through this act of governance, this terrain, together with the green belt, can contribute to the protection of the natural areas, landscapes, forrests, water resources, farmlands, to the promotion of the local eco-economy, as well as to the strengthening of connections between these areas and the city (fig. 6). From an environmental point of view, one of the greatest problems of the city of Timişoara is the quality of the air (The National Agency for the Protection of the Environment, 2011). The number of days in which certain areas of Timişoara are exposed to a concentration of particles in the air high above the European admitted average (PM 10, with a diameter under 10 micrometers) was 136 days in 2008. The admitted values of the pollution with the PM 10 particles is of 40 g/m3 within a year and 50 g/m3 within a day, but not more then 35 times a year. The current situation regarding the pollution in the Growth Pole of Timişoara is the result of two types of factors – interior and exterior to the city. The interior factors are caused by the investement boom, begun after the year 2000, which was based only on real estate profiteering and concentrated on the communist era industrial areas, where the buildings were demolished and the terrain kept unbuilt. Through this policy, to which the local authorities were passive, the investors hoped to obtain a constant raise in the terrain price, but after 2009, the speculative bussiness in real estate suddenly ended and, as a result, in Timişoara's city center there are now approx. 100 ha of unbuilt terrain, phenomenon that affects the quality of the air. The exterior factors take into consideration the winds that blow from the North-West (13%) and from the West (9,8%) bringing dust from the fields of Panonia. This phenomenon has been aggravated by the merging of traditional agricultural lots, which are rather long and narrow, a prerequisite of intensive agriculture, thus forming large plots in different growth stages. The green spaces ("the cities' lungs") balance temperatures and purify the air. Thus, a hectar of vegetation/forrest gives 220 kg of oxygen daily, consuming at the same time 280 kg of carbon dioxide and retaining 50% of the atmospherical dust (The National Agency for the Protection of the Environment, 2011). Following our proposals, between 2001-2009 a 100 m wide protection forrest was planted in the administrative territory of Timişoara, towards the Nort-West, occupying a surface of 50 ha. This first measure must be accompanied by a much more complex one that should concentrate on unifying the previously mentioned natural protected ares, which are situated towards North-West, by a green corridor irrespective of the auto belt (fig. 6).

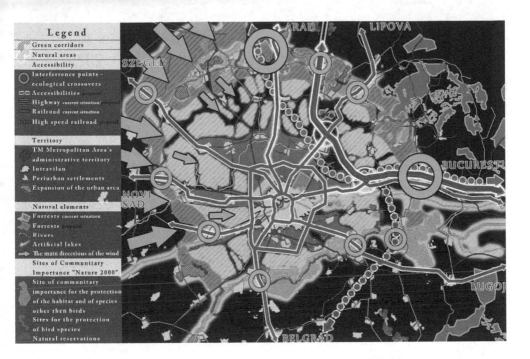

Fig. 6. The Behaviour Setting of the Growth Pole Timişoara

Only by strengthening the polycentric development and inovating the network of metropolitan areas through cooperation with rural and peripheral settlements, through new forms of governance, partnership between rural and urban areas and drafting regional and subregional development strategies can the Behaviour Setting of the Growth Pole Timişoara enter the European competition. At this level of our study, besides Timişoara's expansion towards the North, due to the attraction between Timişoara and Arad, there can also be noticed an expansion towards the East, near the Euroregional "Traian Vuia" Airport of Timişoara, which could become a multi-modal transportation hub that will allow for the optimization of urban logistics. In this hub the future urban train, that will cross Timişoara (and that can continue up to Arad) meets the airport, highway and possibly the high speed railway (300 km/h), which should, in this case, have a station in this point (fig. 7). By doing so, the central area of Timişoara can be freed from the existing railway barrier. The participation of the citizens from a Behaviour Setting to the decisional proccess implies their involvement in all stages of territorial and urban planning. This cannot be achieved whithout a prolonged democratical exercise, while currently in Romania this participation is only formal (an imposed requirement to the authorities by the European norms). As a consequence, in this transition stage towards a consolidated democracy, the position of our Research Group for Sustainable Territorial Development is concentrated on supporting of the citizens' rights to be informed by data transparency, involved and have access to justice.

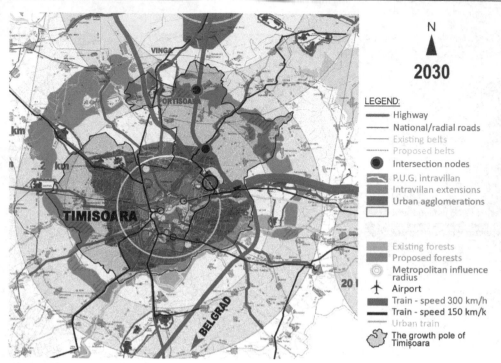

Fig. 7. The Behaviour Setting of the Growth Pole Timişoara

2.6 The relationship between the Behavior Setting of Timişoara City and good governance

The next level of our study deals with the good governance of the Behaviour Setting of Timişoara City. It has a surface of 13.003,87 ha, out of which 6870,21 ha are intravillan. The city population is 334.089 inhabitants (including 16.438 commorants, of which the majority are students), according to the 2002 census, with a decrease in population of 14% in comparison to 1990. In the residential area (2643,74 ha - 53,15% of the intravillan area), the average is 2,2 rooms/housing unit, with a density of 367,70 housing units/1.000 inhabitants. The overall density in the existing intravillan is 49,1 inhabitants/ha, while the average residential areas' density is 126,37 inhabitants/ha. The inhabitancy index is 13,1 sqm inhabitable surface/inhabitant. From the total of 122.195 housing units, 71,30% are in condominium type buildings, while 28,70% are individual housing units (fig. 8).

Hopefully, through good governance, a density of 58,22 inhabitants/ha will be reached, an aspect of whose neccesity will be further developed in our studies. This low density (fig. 9), in comparison to the densities in Europe, makes it difficult for certain facilities to survive without substantial subventions. Our study Master plan for the densification of the urban pattern in Timisoara City (Radoslav et al., 2009), presents a series of measures for each Neighbourhood, Vicinity Unit and Block, to reach a density of 25 housing units/ha in the individual housing units areas (in comparison to the existing density of 3-5 housing units/ha), and 250 housing units/ha in the condominiums areas (in comparison to the

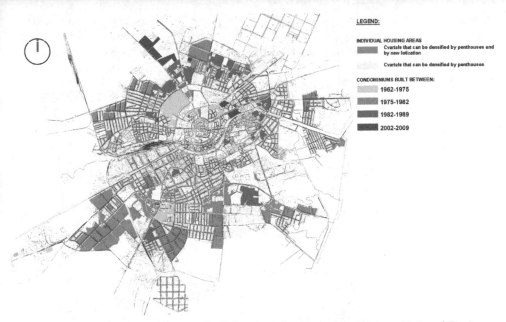

LEGEND:

INDIVIDUAL HOUSING AREAS

Cvartals that can be densified by penthouses and by new lotization

Cvartals that can be densified by penthouses

CONDOMINIUMS BUILT BETWEEN:

1962-1975

1975-1982

1982-1989

2002-2009

Fig. 8. Typology of housing units in the Behaviour Settings of the Vicinity Units of Timișoara City (condominium and individual housing units)

The density of the population across the entire teritory of the city (housing units/ha)

cca.300/1ha

cca.200/1ha

cca.150/1ha

cca.100/1ha

cca.75/1ha

cca.50/1ha

cca.25/1ha

cca.15/1ha

cca.3/1ha

Fig. 9. The density of the population across the territory of the Behaviour Settings of the Neighbourhoods of Timișoara City

existing 300 housing units/ha), by transfroming the housing units on the ground floors into service areas. The proposed measures deal with avoiding the deepening of the social polarization and of the potential risk of social fragmentation, stimulating the labour market, reduction of school dropout rates, as well as of the risk of underpriviliged districts formation in already vulnerable areas. Timişoara green areas' surfce is 510 ha, out of which parks 117,57 ha, green squares 21,58 ha and Pădurea Verde forrest 50,7 ha. 84% of this surface is actually occupied by vegetation. Even though in 2005 the green surface represented over 16 sqm/inhabitant it was poorly distributed, as the green spaces were concentrated in the city center (fig. 10). It is therefore neccessary to create a park in the South-East part of the city. Consistent with the approved national document no. 114/2007, by the end of 2010 the green surface should have reached 20 sqm/inhabitant and should reach 26 sqm/inhabitant by 2013.

Fig. 10. The current situation of the green areas and of the schools and sports facilities in the Behaviour Settings of the Neighbourhoods of Timişoara City

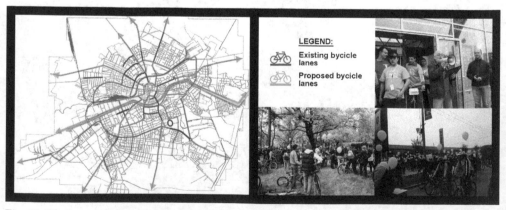

Fig. 11. The current situation and our proposal regarding bycicle lanes of the Behaviour Setting of Timişoara City, which benefits from the support of the population

In order to accomplish this basic need, the local administration should buy approx. 60 ha, out of which 30 ha for parks should be within the city (in the Romanian legislation, a park must have over 10.000 sqm). For the rest of the green spaces, such as a network of green squares in every Neighbourhood, 300-500 m at most from any housing unit, a series a measures should be provided at the level of each Behaviour Setting of the Neighbourhood. Another measure reffers to the realization of a major non-motorized transportation network (bycicle lanes, fig. 11) in order to reduce pollution, which is based on the philosophy of a system of radials up to the ring no. 2, connected through rings, as well as of a system of by-passes alongside the Bega Canal and the railroad, which will complete the current, not yet structured, network. All of these measures are currently in process of implementation, due to the pressure of the NGOs.

3. The relationship between the Behaviour Settings of the subunits and good governance

The hierarchy cannot stop here because local governance (through the previously mentioned subsidiarity and procesuality principles) must reach the level of a group of inhabitants, passing through District, Neighbourhood, Vicinity Unit, Block, Group of Housing Units and the units around a Condominium Building's Staircase. From here onward, the involvement of the citizens as an active part in the decisional process must be even stronger, so that every spatial unit will become a Behaviour Setting. This implies the existence of a District Council (a consultative non-juridical entity), in every District in the city, of an association at the level of every Neighbourhood, Vicinity Unit and Block, as well as encouraging the already existing Owners' Associations (juridical entities), at the level of every Staircase of a Condominium Building. These councils and associations should have a very important role in the decisional process regarding their spatial unit, as well as in the distribution of money from the taxes cashed in by the States' Budget (in particular, the 40-45% that form the local budget), which should be redistributed to each and every level mentioned above. Only this way can the Toledo Declaration, that supports *"an implication, a taking on tasks and a responsabilization of the factors, at multiple levels and from an integrative point of view"* be applied. Governance should now take place, on a local level, through the Decisions of the Local Council (Radoslav, 2000), which should be based on participation and support of every organization from every Behaviour Setting that is affected by that decision.

3.1 The relationship between the Behaviour Setting of a District and good governance

With the inscrease in size of some cities, a new unit, namely the Behaviour Setting of a District, with its own rules regarding governance, appeared in order to provide better administration. But this Behaviour Setting can be composed of very different spatial and social dimensions (in Bucharest, the capital city of Romania, it can have hundreds of thousands of inhabitants, in Timişoara – tens of thousands of inhabitants, while in Lugoj, a smaller city near Timişoara, the Behaviour Setting of a District can have a few thousands of inhabitants).

In Timişoara 18 such Districts Behaviour Settings function, based on the city's tradition (fig. 12). Problems become even more complicated as these districts are, most of the times, composed by more then one social community, with a certain identity, life style, human

Fig. 12. The Behaviour Settings of the Districts of Timişoara City

behaviour and different human needs, which require specific rules. The defining characteristics of a district are its accessibility to parks (green spaces larger than 10.000 sqm), the main transportation and bycicle lanes' networks, public spaces – the major public square and the district's promenade, highschools, swimming pools, the district's health and major commercial networks and the sacred places (eclesiastical buildings and cemeteries). The catholic confession has placed its churches in the historical areas of the city, the orthodox confession has positioned its churches depending on Neighbourhoods, while the neoprotestants placed their churches depending on the Vicinity Units – fig. 14). However, these Districts are not always perfectly divided into smaller units, with identifiable social units (one such subunit could belong to two different Districts), due to the interventions which influenced their development in the communist era, and this leads to the improper functioning of the Behaviour Settings of the Districts.

One of the major dysfunctionalities of these subunits is the presence of the main traffic roads inside the Behaviour Settings. This became an even bigger problem during the last years, because of the increase in the number of autovehicles that populate the streets of Timişoara. Thus, in comparison to a total number of 239.924 autovehicles in 1997, the 2011 traffic comprised 368.248 autovehicles, which indicates an increase in traffic with more then 53% during the last 14 years (Urbanistic General Plan for Timişoara, 2011). An element that can contribute to the good governance of the Behaviour Setting of the District through a stronger involvement are the District Councils, which have a consultative character and have functioned in Timişoara since 2000, but they did not always take into consideration the point of view of the associations that represent lower subunits, influenced by that decision.

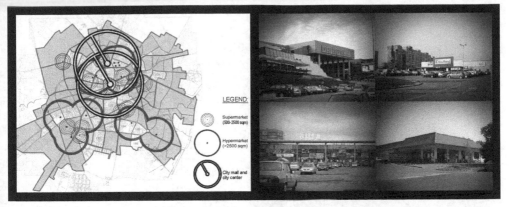

Fig. 13. The main commercial network of the Behaviour Setting of Timișoara City

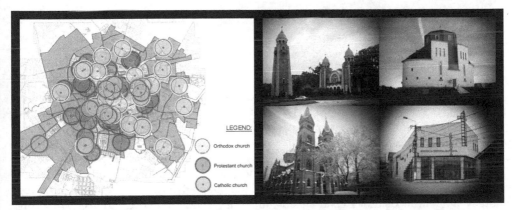

Fig. 14. The churches of the Behaviour Settings of the Districts of Timișoara City

3.2 The relationship between the Behaviour Setting of a Neighbourhood and good governance

Our studies have introduced new levels of governance, such as the Neighbourhood, Vicinity Unit, Block or the Group of Housing Units. We have focused on the Neighbourhood, which we consider to be a Basic Spatial Unit for a Behaviour Setting, as well as on its delimitations and on the methods that allow the introduction of the needed functions, through good governance. The Neighbourhood proposed by us has a social dimension of approx. 5.000-10.000 inhabitants (Alexander, 1977) and a physical one of approx. 350x750 m, which represents an area of approx. 7-40 ha. These Behaviour Settings have some common overall characteristics that offer them an identity, such as a landscape with similar characteristics, a representative historical evolution in a certain period of time, population with homogenous structure, similarity in the shape and areas of the lots and buildings, of occupying the terrain or a similar legislative regime of the properties, homogeneous urban rules regarding the allowed functions, etc.

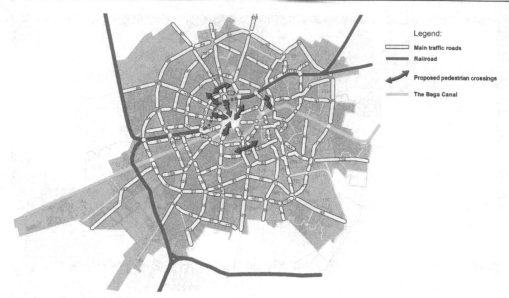

Fig. 15. The main traffic roads network which delimit the Behaviour Settings of the Neighbourhoods of Timişoara City

The delimitations between the Neighbourhoods have been determined by major natural barriers, such as the Bega River, or by built elements, such as streets with major traffic or the railroad, that currently crosses Timişoara City's central area. Thus, major traffic is not admitted within the limits of a Neighbourhood, where the speed limit is 30 km/h. In Timisoara, over 100 such Neighbourhoods have been delimitated, out of which more than 70 have as main function residential. In these particular Neighbourhoods, bicycle and pedestrian traffic have priority. If the criteria of the population number cannot be achieved, more Neighbourhoods can be connected by under and overground pedestrian crossings (fig. 15). From our studies regarding this type of situations, we present a solution proposed for the Northern part of Timişoara City, as shown in the fig. 16. The analysis of each Neighbourhood was based on many criteria, out of which we mention a Neighbourhood's connectivity to the public transportation system, as well as the network of public stations, provided with attractive functions and activities and situated no more than 300-500 m from any housing unit (fig. 17). The study of this pattern proves that there are some peripheral Neighbourhoods that do not have access to the public transport system, because their density does not reach 25 inhabitants/ha (fig. 9), minimum prerequisite for the efficiency of public transport. This leads to the inability to support a public transport which is cheap and accessible to everybody, especially in the peripheral neighbourhoods, where it should play a key role in order to diminish the physical isolation of these neighbourhoods.

Another important pattern which we analysed for these Behaviour Settings is the access to green squares (green spaces of approx. 5.000 sqm), so that no housing unit is more than 300-500 m from this facility. One can observe that more than 50% of the Behaviour Settings of Neighbourhoods of our city do not respect this criterion, but for this pattern, the citizens' participation in promoting it is very important (fig. 10). These green squares must

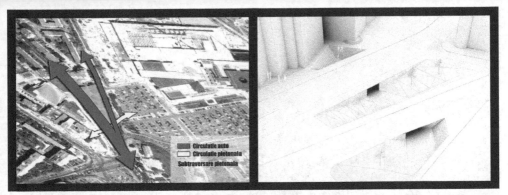

Fig. 16. Proposal for a pedetsrian underground crossing that unites the Behaviour Settings of two Neighbourhoods in the Northern part of Timişoara City

Fig. 17. The public transport network and the bus stops in the Behaviour Settings of the Neighbourhoods of Timişoara City

eventually form a green network, which should be uniform throughout the city, together with the district parks and the green riverside of the Bega Canal. Another important pattern is acces to water, that could be achieved, in the years to come, through the rehabilitation of the Bega Canal (fig. 18), thus becoming an important axis on the East-West direction. Besides this major operation, the 20 ha of water canals that exist on the outskirts of Timişoara City must also be emphasized. These proposals regarding the green and blue spaces are merely trials that sustain the closing of the metabolic urban cycles at a local level.

The need to ensure an equal opportunity for education, as well as proffesional training oriented towards maket demands and inclusion leads to the conclusion that in every Neighbourhood there should be a highschool, with sports grounds and other sports facilities, situated at no more than 300-500 m from every housing unit (fig. 10), accessible without crossing a major traffic road. Our studies show that the problems regarding this pattern are not entirely solved, especially the ones concerning the sports grounds and other sports facilities. Without good governance, this situation can lead to social fragmentation and massive school dropout rates. For this level, the participation of the population is also

Fig. 18. The riverside of the Bega Canal – current situation and proposal from disertation papers within the Master of Urbanism and Territorial Planning, coordinated by The Reseach Group for a Sustainable Territorial Development

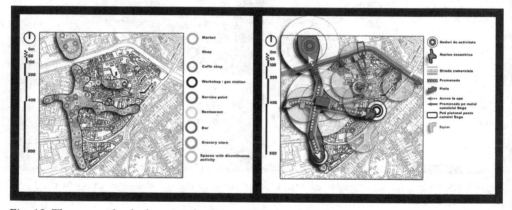

Fig. 19. The network of education facilities, services, commerce and other public facilities in the Behaviour Setting of the traditional Fabric Neighbourhood from disertation papers within the Master of Urbanism and Territorial Planning

very important, in order to encourage the buying and afterwards the maintaining of the terrains needed for these facilities, with municipal funds and funds belonging to the Behaviour Setting of that certain Neighbourhood. Obviously the main commercial network (fig. 13), a result of the consumer society, is well represented. A very important factor in supporting the eco-products of the Growth Pole Timişoara is the creation of farmer markets in every Neighbourhood, but the local community of each Neighbourhood should provide some measures, through approved rules, that will maintain the services and commerce in the area (fig 19), which is very affected by the aggressive presence of the multi-national corporations' chains of stores. The public spaces network (Radoslav & Cosoroaba-Stanciu, 2010; fig. 20), that makes the life within a community more animanted and healthier (Gehl, 2011), is very similar to the one existing in the year 1900, when Timişoara City's population was half of what it is today.

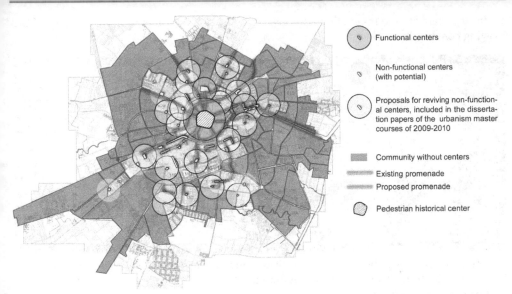

Functional centers

Non-functional centers
(with potential)

Proposals for reviving non-functional centers, included in the disserta-
tion papers of the urbanism master
courses of 2009-2010

Community without centers

Existing promenade

Proposed promenade

Pedestrian historical center

Fig. 20. The public spaces network of the Behaviour Settings of the Neighbourhoods of Timișoara City (current situation and proposal)

The disertation papers within the Master of Urbanism and Territorial Development, "Politehnica" University of Timișoara, produced 50 such proposals, out of which we present a few (fig. 21). These projects raise the attractivity of the Behaviour Setting of the Neighbourhoods, encourage inhabitants to identify with the place and the strengthening of democracy, co-existance, changes, civic progress, diversity and, last but not least, freedom, both indiviual and collective, which are key elements of the European spirit. We also mention that the studied patterns of the health network, the working community network, etc., have begun to naturally coagulate in clusters, action that must be continued through specific measures, all part of our strategies of regenerating the urban economy (fig. 19). One of the conclusions of these studies was that the Neighbourhoods that do not have a reasonable density (fig. 9) cannot sustain neither one of these functions, needed for a Behaviour Setting to function correctly, a school, sports fields, a green square, public transport, salubrity, etc., without substantial financial support from the larger whole. In Timișoara, there are some peripheral Neighbourhoods, that resulted either by the absorbtion of villages with agricultural activities over time, or by new sprawl-type developments, with a density of approx. 5-10 housing units/ha in this situation. In conclusion, these areas should urgently and compulsory go through a densification process, that should not affect the current value of the properties, nor the identity of the place (as it happened in the last few years, through insertion of multi-leveled condominium buildings in areas with a low story limit, without the neighbours consent). A second conclusion is that the Behaviour Setting of Vecinity Units in Timișoara City, together with the Behaviour Setting of every Neighbourhood, should have *"a powerful control of the available terrain and of the speculative development"* (Toledo Declaration, 2010), especially regarding lacking public facilities (green spaces, education facilities). The Master Plan for the Densification of the Urban Pattern in Timisoara (Radoslav et al., 2009) proposes the solving of these problems after having a series of discussions with the population and only by taking into consideration the identity of the existing communities.

Fig. 21. Proposals for public spaces in the Behaviour Setting of the Neighbourhoods of Timişoara City from disertation papers within the Master of Urbanism and Territorial Planning

3.3 The relationship between the Behaviour Setting of a Vicinity Unit and good governance

The following level of the study reffers to the good governance of the Behaviour Setting of a Vicinity Unit, with a social dimension of approx. 500-1.500 inhabitants; thus, more Vicinity Units compose a Neighbourhood. In Timişoara, 248 Vicinty Units have been identified (fig. 22), for which building regulations have been established (Radoslav et al., 2009), in such a way that the identity of the place will not be destroyed and the existing value of the properties will not be diminished. These rules have been very important for the population in those areas, since it helped them get used to having access to justice, as, in their race towards profit, the developers have built, after 2005, a series of condominiums in the areas of individual housing units, often without the neighbours consent and without abiding regulations regarding the minimun distances needed for natural daylight and intimacy. The regulations regarding the allowed plot occupancy, land use and densities have been drastically ignored.

An operation that has lowered the quality of life in the Vicinity Units was the proccess of adding attics to condominium buildings in the areas with a density of over 300 housing unints/ha, which lead to even higher densities and emphasized the absence of the necessary facilities such as green spaces, kindergardens, schools, sports grounds, parking, etc. The programme presented in the Master Plan for the Densification of the Urban Pattern in Timişoara studies (Radoslav et al., 2009) was an attempt to stop these errors by establishing some rules for each Vicinity Unit that should respect the technical norms. As an example of good governance, one can observe the programme that begun in the year 2000 regarding the playgrounds for children, which is almost entirely completed (fig. 24), as well as the one concerning public water fountains, a network that is present in every Vicinity Unit (fig. 25).

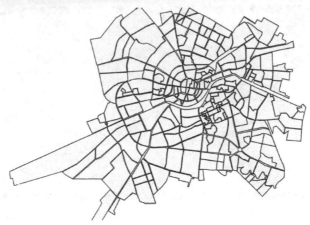

Fig. 22. The Behaviour Settings of the Vicinity Units of Timişoara City

Fig. 23. Commercial spaces, created through appartment refunctionalizing on the ground floor of condominium buildings, in the Behaviour Settings of the Vicinity Units of Timişoara City

Fig. 24. The playground network in the Behaviour Settings of the Vicinity Units of Timişoara City and examples of such playgrounds

Fig. 25. Existing public facilities in the Behaviour Settings of the Vicinity Units of Timişoara City, public fountains and water pumps, and lacking functions (planned parking spaces)

3.4 The relationship between the Behaviour Setting of a Block and good governance

The next level of the study refers to the good governance of a Block, with a social dimension of approx. 100-500 inhabitants; several Blocks can form a Vicinity Unit, and within the limits of a Block traffic is completely forbidden. Quality of life is also determined by the acceptance of the idea that people try to give the place they live in a personality of its own (Alexander, 1977; fig. 27). Unless we offer inhabitants the facilities they need, they tend to use the public terrain according to their own rules, which prevent the area from developing harmoniously – the green areas are occupied by vehicles (fig. 25). Good governance implies rules established through Decisions of the Local Council, regarding the terrain in front of the condominium buildings, which should be provided with playgrounds, sitting areas for elders or commerce (fig. 28). For each of the 1089 of Timişoara's Blocks, there have been established some measures and rules that encourage social life preventing these Blocks from becoming underprivileged areas.

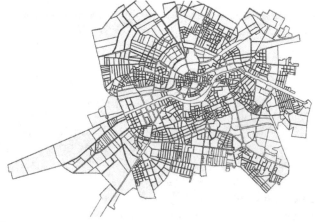

Fig. 26. The Behaviour Settings of the Blocks of Timişoara City

Fig. 27. Modified balconies in the Behaviour Settings of the Blocks of Timișoara City

3.5 The relationship between the Behaviour Setting of a Group of Residential Units and good governance

The last level of the study refers to the good governance of the Behaviour Setting of a Group of Residential Units, which is divided according to the two types of residential units, namely individual housing units and collective housing units (the units around an Apartments Building's Staircase); this Behaviour Setting has a social dimension of approx. 30-100 inhabitants. The people's discontent with condominium-type projects, focused only on economical efficiency maximization, is obvious, as are their attempts to adapt these buildings to their needs by their own means. For good governance, it is compulsory to accept the rules of organic growth (Alexander, 1977) in order to maintain the health of these Behaviour Settings, thus resulting intervention methods that should take into consideration the interests of the members of the corresponding social groups. These needs can be found in our proposals, which are approved by Decisions of the Local Council, such as the need for an exterior public room in front of a condominium building, ground floor modified balconies (fig. 28), direct access from the apartments towards the green space situated in

Fig. 28. Exterior public rooms in front of condominium buildings, modified groud floor balconies, elder in the park and improvised green areas in the Behaviour Setting of a Group of Residential Units of Timișoara City

front of the building, diverse facades and balconies, added roofs, the thermic isolation of the facades, as well as other operations that imply exemption from local taxes.

4. Conclusion

"The urban regeneration" and "the integrated approach" require a new **"urban alliance"**, shared by all actors involved in the proccess of "building the city": the owners, the finances, the inhabitants, the public authorities, the experts, etc., at all the levels of Behaviour Setting mentioned above. This new "urban alliance" should be based on consensus and it should be legitimized by new forms of governance, in which the social networks play a very important role. The public financing for urban regeneration is the engine that attracts private funds, which should join the Public-Private Partnerships. Thus, the public budget, that consists of 40% of the taxes which remain at the disposal of the local authorities, should be further redistributed, according to the principles of subsidiarity and procesuality: 40% should remain at the disposal of the local authorities, 30% should go to Districts' authorities, 20% should go to the Neighbourhoods, while 10% should go to the Vicinity Units. Besides these, financial stimulents should be created, as well as tax exemption for private companies, thus raising the involvement of the private domain, financial agents and other urban actors in the urban regeneration. The purpose of this proposal is the strenghtening of good governance at every level, that should be directed towards **revalorification, recuperation** and **reinventing of the "existing city"**, thus optimizing the human, social, material, cultural and economic capital, which has developed throughout the history, as well as using these elements in order to build efficient, inovating, inteligent, more durable and socially integrated cities (Toledo Declaration, 2011). The spread of internet use from the early '90s has favoured the formation of a new society, which, at first sight, seems not to take into consideration spatial limits, in which the public-private relationship seems to be destroyed and which seems to anihilate the specificity of local communities. This situation has peaked at the begining of the 21st century, with the spreading, at global level, of the 2.0 web, or of the "web of social interaction". Social networks, such as Facebook, Twitter or hi5, as well as blogs occupy an increasing role in the private lives of the internet-users world over. Our opinion is that by facilitating access to information through these methods, the social networks strengthen the cooperation of the inhabitants for the satisfaction of the human needs (Maslow, 1987) in every Behaviour Setting, through the submination of the control of information at a central level. Thus, the Behaviour Settings are no longer enclosed, which makes living in them much more pleasent, according to the analysis on a Behaviour Setting of a Neighbourhood from Timişoara, coordinated by R. Radoslav (Isopescu et al., 2009) and presented at the 2009 Rotterdam Biennale of Architecture. This situation leads to a new type of public debate, both on a horizontal, as well as on a vertical level. This type of functioning is very much alike to the one in which Europe, the Euroregion, the Region, County, Growth Pole, City, District, Neighbourhood, Vicinity Unit and Block are organised in a network, as parts of a larger whole. In case good governance is not applied in time, the connections between the leaders and the citizens can be broken; this phenomenon is already present, with high intensity, in almost every European city, and especially in the Romanian ones. Our developing studies are currently concentrating on the qualitative evaluation (with total, partial or zero satisfaction) regarding a variety of patterns reffering to the human scale.

5. Acknowledgement

We thank: IGEAT – Institut de Gestionde l'Environnement et d'Aménagement du Territorie, Université Libre de Bruxelles, for the 2006 ESPON project 3.2. Spatial Scenarios and orientations in relation to the ESDP and Cohesion Policy (Final Report)

The strategic grant POSDRU 107/1.5/S/77265, inside POSDRU Romania 2007-2013 co-financed by the European Social Fund – Investing in People.

6. References

Alexander, C.; Ishikawa, S. & Silverstein, M. with Jacobson, M.; Fiksdahl-King, I. & Angel, S. (1977). *A Pattern Language. Towns. Buildings. Construction*, Oxford University Press, ISBN 0-19-501919-9, New York

Alexander, C.; Hajo, N. & Artemis, A. (1987). *A New Theory of Urban Design*, Oxford University Press, ISBN 0-19-503753-7, New York

Butters, C. (2004). A Holistic Method of Evaluating Sustainability, Building and Urban Development in Norway, In: *www.universell+untforming.miljo.no*, 14.07.2011, Available from:
<http://www.universell-utforming.miljo.no/file_upload/idebank%20article%20chris%20butters.pdf>

Gehl, J. (2011). *Viața dintre clădiri – Utilizările spațiului public*, Editura Igloo Media, ISBN 978-606-8026-12-1, Bucharest

German EU Presidency & German Federal Ministry of Transport, Building and Urban Affairs (BMVBS) (2007). Leipzig Charter on Sustainable European Cities, In: *www.eukn.org*, 13.07.2011, Available from:
<http://www.eukn.org/E_library/Urban_Policy/Leipzig_Charter_on_Sustainable_European_Cities>

IGEAT – Institut de Gestionde l'Environnement et d'Amenagement du Territorie, Universite Libre de Bruxelles (2006). ESPON project 3.2. Spatial Scenarios and orientations in relation to the ESDP and Cohesion Policy. Final Report, In: *www.espon.eu*, 14.07.2011, Available from:
<http://www.espon.eu/export/sites/default/Documents/projects/ESPON2006P rojects/CoordinatingCrossThematicProjects/Scenarios/fr-3.2_final-report_vol1.pdf>

Isopescu, B.; Piscoi, C.; Rigler, R.; Sabău, S.; Sgîrcea, M.; Spiridon, A.; Tomescu, A. & Văleanu, P.; coordinated by Radoslav, R. (2009). *inBetween*, Editura Politehnica, ISBN 978-973-625-939-5, Timișoara

Lang, J. (1994). *Urban Design – The American Experience*, John Wiley & Sons, ISBN 0-471-28542-0, New York

Maslow, A. (1987) *Motivation and Personality* (3rd edition, revised by Frager, R.; Fadiman, J.; McReynold, C. & Cox, R.), Harper & Row, ISBN 978-006-041-987-5, New York

Radoslav, R. (2000). *Topos comportamental – Armonizarea dintre spatial urban si comportametul uman*, Editura Marineasa, ISBN 973- 9485-71-5, Timisoara

Radoslav, R.; Anghel, A. & Branea, A. (2009). Densification of Singular Housing Neighbourhoods. *Scientific Bulletin of the "Politechnica" University of Timișoara, Romania*, Tomul 54(68), Fascicola 1, (2009), pp. 37-41, ISSN 1224-6026

Radoslav, R.; Branea, A.; Gaman, M. & Morar, T. (2010). Master Plan for the Densification of Urban Fabric in Timisoara. *Urbanismul Serie Nouă*, No. 4, (April 2010), pp. 74-77, ISSN 1844-802X

Radoslav, R.; Branea, A.; Badescu, S.; Gaman, M.; Morar, T. & Nicolau, I. (2010). *Organic Growth. Studies for Territorial Development, Urbanism and Urban Design*, Editura Orizonturi Universitare, ISBN 978-973-638-440-0, Timisoara

Radoslav, R. & Cosoroaba-Stanciu, E.. (2010). The Renewal of Public Spaces as an Instrument of Urban Regeneration in Local Communities of Timisoara, *Selected Topics in Energy, Environment, Sustainable Development and Landscaping, 6th WSEAS International Conference on Energy, Environment, Ecosystems & Sustainable Development (EEESD 10), 3th WSEAS International Conference on Landscape Architecture (LA 10)*, ISSN 1792-5924, ISSN 1972-5940, ISBN 978-960-474-237-0

Saunders, W.S. (2005). *Sprawl and Suburbia*, University of Minnesota Press Minneapolis, ISSN 0-8166-4755-0, London

Spanish EU Presidency (2010). Toledo Informal Ministerial Meeting of Urban Development Declaration, In: *www.eukn.ro*, 13.07.2011, Available from: <http://www.eukn.org/News/2010/June/Ministers_of_Housing_and_Urban_De velopment_approve_the_Toledo_Declaration>

The National Agency for the Protection of the Environment (2011). Calitatea aerului ambiental. Raportarea anuală, In: *www.arpmtm.anpm.ro*, 13.07.2011, Available from: <http://arpmtm.anpm.ro/articole/raportarea_anuala-51>

The Regional Development Agency for the Western Region (2011). Regiunea Vest, In: *www.adrvest.ro*, 13.07.2011, Available from: <http://www.adrvest.ro/index.php?page=domain&did=47&maindomain=yes>

The Romanian National Prognosis Comitee (2011). Prognoze, In: *www.cnp.ro*, 13.07.2011, Available from: <http://www.cnp.ro/ro/prognoze>

Urbanistic General Plan for Timişoara (2011) Fase 1. Fundamental studies – analysis and diagnosis, In: *www.pugtm.ro*, 21.09.2011, Available from: <http://www.primariatm.ro/uploads/files/PUG/TRAFIC/parte%20scrisa>

3

Sustainable Urban Design and Walkable Neighborhoods

Theresa Glanz, Yunwoo Nam and Zhenghong Tang

University of Nebraska-Lincoln

USA

1. Introduction

Urban development within the United States has not remained stagnant as evident by the development patterns that have evolved over time. When urban development was beginning in the United States there was a mix of land-uses which were necessary due to the limited transportation options available at the beginning of the twentieth century and before. Sustainability was related to self-preservation and was partially focused on the ability to get to the needed destination which was accomplished through use of one of the following available transportation modes; horse, trolley, train, and/or walking. A close proximity to the frequented locations was highly desirable due to the limited range of these transportation modes. However, as the evolution of the automobile occurred and became more attainable by households, urban development began a transformation that would help push housing away from the city center and away from desired destinations such as places of employment, shopping, and school. By the mid 1900s, the private automobile was becoming the primary mode of transportation for households and cities would begin tailoring infrastructure to accommodate the increasing numbers of automobiles in use. Sustainability during the height of suburban neighborhood development has been related to personal space preservation and has had little to do with public transportation, environmental preservation, and household finances.

For middle income families in the United States this reliance on the automobile coupled with living in the suburbs would not become a major financial hardship until the beginning of the twenty-first century when fuel prices would dramatically increase in a short period of time. Based on the U.S. Energy Information Administration website prior to the beginning of the 2005 hurricane season (which runs June through November of each year) the average monthly retail prices for gasoline in the United States Midwest region were consistently below $2.00 per gallon. Beginning with the 2005 hurricane season, fuel prices would progressively increase until the average Midwest retail price reached a monthly average high of $3.99 per gallon in June 2008 (other regions were higher such as the western state of California where fuel prices averaged $4.48 per gallon). Based on the Bureau of Labor Statistics Consumer Expenditure Survey the annual cost of gasoline and motor oil expenditures would rise 69.9% between the years 2004 and 2008; during this same time period the median household income in the United States would remain stagnate. Had the "ideal" suburban home and the need to own a car to commute to and from the suburbs become a unsustainable reality for many households?

This chapter discusses how walkable neighborhoods contribute to the goal of sustainable communities. The topics covered are the history of neighborhood development, defining walkability and measurement tools, and the application of walkability principles into new developments and incorporating walkability into redevelopment projects. The first section provides an overview of neighborhood development in the United States and incorporates such ideas as presented through Clarence Perry's Neighborhood Unit design through the current movement of New Urbanism. The second section explains what walkability is and the elements to consider when trying to assess the environmental qualities that contribute to walkability. The following sections focus on the principles being used in new urban developments that encourage walking and include a case study. The final section discusses opportunities and actions needed to incorporate walkability in existing neighborhoods.

2. Neighborhood development patterns

Within the United States urban neighborhoods can typically be classified into three distinct development types each representing a different attitude towards the mixing of land uses as well has each having different emphasis on the importance of the automobile; the three neighborhood types being discussed are traditional, conventional, and New Urbanism. Traditional neighborhoods were the prevailing type of urban development prior to World War II, conventional neighborhoods flourished during the years following World War II and New Urbanism is a relatively recent design movement that is a response to the sprawl created by the conventional suburban neighbourhood and derives its design elements from the early pre-suburban neighborhoods of the inner city.

Traditional neighborhoods are those neighborhoods built during the first or second ring of development in an urban setting within the United States. Living in these neighborhoods meant that a person lived in or close to the city center where he/she could easily walk to their intended destination which was necessary as the automobile was not widely used or owned by American households prior to the mid 1900s. Traditional neighborhoods are characterized by streets that are laid out on the grid system, close proximity to the city center where there may be a mix of land uses, higher population density, and buildings that are set relatively close to each other due to the smaller lot sizes. These neighborhoods may have historically been serviced by public transportation such as trolleys due to the higher population density which could help subsidize the transportation system. In addition to allowing for easy mobility, there were a multitude of accidental and intentional socialization opportunities due to the tightly built and mixed-use environment. Within these inner neighborhoods the residents could find shopping and employment opportunities as well as housing but it was the functionality of the neighborhoods that determined what type of housing was available. To help maximize proximity to destinations, due to lack of transportation options, housing units could be found above stores or to be tightly clustered together, such as row houses.

These early neighborhoods were not without problems. There were issues with substandard housing, lack of open space and crime to name just three. Housing advocates pushed for housing reform to relieve problems with congestion and to reduce the incidents of widespread illness due to the overcrowding and unsanitary conditions often found within the early inner neighborhoods. The quest for housing reform would begin to push housing outward where there could be an increase in space between homes, where open spaces for recreation could be incorporated, and where the number of units per lot could be reduced to one. In the 1920s zoning would begin playing a crucial role in this separation of housing from other neighborhood functions such as employment and shopping. The 1926 Supreme Court case of The Village of Euclid vs. Ambler Realty Company declared that exclusive zoning was not unconstitutional and could be construed as police power in safeguarding against conditions that could be considered detrimental to human health. This case would help set the stage for future exclusively zoned developments in which seemingly all types of land uses would be segregated.

Suburban neighborhoods are a product of The Zoning Enabling Act, housing reform and the post World War II era when people wanted to move away from the congestion and crowding of the inner city areas and home ownership became a driving economic goal for many families. Conventional suburban neighborhoods, which have been the predominant type of residential development since the end of World War II, are often referred to as cookie-cutter developments due to the repetitive exterior designs of the homes which typically feature a prominent drive-way and garage. These neighborhoods are zoned primarily for single family homes and the infrastructure is designed to contain streets that are curvilinear and that may terminate in cul-de-sacs which may not be pedestrian friendly. The automobile dominates the transportation system in the conventional neighborhood as stores, schools, and employment may be outside of a reasonable walking distance and out of the reach of the public transportation system. Additionally, walking in suburban neighborhoods may be limited to leisure walking as accessibility of public transportation may be restricted or non-existent due to the distance from the city center and/or lack of ridership that would support the cost of operating a transportation system to the area. The private automobile truly dictates the street system when the development occurs on the urban fringe.

One of the first recognized suburban neighborhoods in the United States is Levittown New York which was created in the late 1940s. This development was initially built in response to the shortage of available housing for returning veterans and their families. With it's sea of single family homes and curvilinear streets it offered affordable housing in five architectural style choices and by 1951 more than 17,000 homes had been constructed in Levittown and the surrounding areas, according to Levittown Historical Society's website (www.levittownhistoricalsocierty.org). The infrastructure pattern and the repetition of exterior architecture of homes built in Levittown, New York would be repeated again in a second Levittown in Pennsylvania in the 1950s. Levittown is often considered the first built suburban neighborhood in the United States and the street pattern and repetition of architecture used in this early suburban development continues to be found throughout conventional suburban neighborhoods today.

There is an important distinction between the early suburban neighborhoods and the present-day suburban neighborhoods, while both emphasize automobile ownership through the prominent inclusion of garages and driveways that dissect sidewalks, the emphasis on automobile ownership is considerably more prevalent in today's suburban developments. Early developments, such as Levittown, featured homes with a single-stall garage (if there was one) that was set even with the front of the home or was slightly offset back from the front of the home while today's developments commonly feature a three-stall garage with a driveway that is nearly equal in width and which may protrude several feet in

Fig. 1. Levittown, Pennsylvania. Available from:
<http://www.theurbn.com/2010/09/levittown-urban-revitalization>

front of the home's front door. The expanded widths of garages and driveways may lead to an appearance of shorter expanses of sidewalk that are uninterrupted which may give the perception of an unfriendly-pedestrian environment.

A third type of development in the United States is that promoted by the New Urbanist movement which is a response to the sprawl associated with suburban developments. Beginning in the 1980s, a renewed interest in center city living occurred with the regentrification of older inner city neighborhoods. The attraction of the inner city neighborhoods often is the proximity to work, shopping, entertainment, and/or transportation options. In addition to regentrification, the 1980s brought a renewed interest in the creation of neighborhood developments that would incorporate the perceived physical and social dynamics of the pre-suburban neighborhoods which is the basis behind New Urbanism. The New Urbanist developments work to create compatible mixed land uses that do not require segregation of residential spaces from all working and shopping and which encourages walking as a mode of transportation.

Other names associated with New Urbanism are Neo-traditional Neighborhoods or Traditional Neighborhood Developments which are a modern take on the older inner city neighborhoods of mixed uses and increase land development density and Transit-oriented Developments which strive to create communities that are centered on public transportation with the ability to walk to the transit stations. Unlike the early twentieth century mixed-use neighborhoods and the suburbs of post World War II, the focus of New Urbanism's traditional neighborhood development is not primarily on functionality or housing types but is a shift to a more balanced view that neighborhood functionality and housing can be brought together to create a sustainable, livable community.

To create the desired built environment New Urbanism's design principles includes the use of a grid or undulating street system to maximize pedestrian connectivity, incorporates a mix of compatible land uses that includes housing, retail, and public facilities, and works to create a streetscape that encourages human interaction through the incorporation of design elements such as street furniture and architectural details (front porches on residential units) that extend the living space outside (Rohe, 2009) which is different from conventional suburban neighborhoods that are zoned for a single land use and separation of neighbors.

With it's design departure from the standard suburban neighborhood development within the United States, New Urbanism has been a catalyst for reexamining how neighborhoods should function and the impact of design on functionality; the result is a push for and an ongoing discussion of what makes a neighborhood that is economically, socially, and environmental sustainable. One of the ways that New Urbanism promotes sustainability is through the incorporation of a pedestrian-friendly environment which allows residences to be able to choose walking as an alternative mode of transportation when moving about within the development. This is done by offering a variety of destinations to walk to and incorporates streetscapes that encourage walking, in part, through the reduction in sidewalk breaks by placing the garage and driveway behind the house which is accessible by an alley. By removing the garage and driveway from the front of the house it creates a longer continuous span of sidewalk which made lead to the perception that it is safer to walk.

Fig. 2. Suburban Development (left picture) and Traditional Neighborhood Development (right picture) of Lincoln, Nebraska. Left picture shows suburban neighborhood with garages that protrude beyond the front of the house with driveways that intersect sidewalks. Right picture shows a New Urbanist neighborhood with garage access to the rear of the home and unobstructed sidewalks.

While New Urbanism is a fairly recent concept it is not the first attempt at creating sustainable, equitable neighborhoods that work to balance social and environmental equity. A well-known development attempt at creating a sustainable neighborhood was introduced in 1929 by Clarence Arthur Perry who introduced his plans for the Neighborhood Unit. The features of the Neighborhood Unit had at its core public space that included schools, churches, and open space for recreation. The distance each resident had to travel to reach the core or perimeter commercial space was important and was to be no longer than a quarter-mile walk. The types of streets used within the development were also regulated so that the main arterial streets were along the perimeter which allowed for residents to walk with less fear of traffic (Lawhon, 2009).

Another pre-New Urbanist design was that conceived by Ebenezer Howard. In the late 1800s Ebenezer Howard's vision was the Garden City which was intended to be a system of self-sufficient satellite cities connected to each other via a rail line (Daniels, 2009). The intent of the Garden City (located in the United Kingdom) was to alleviate some of the problems of social injustice found in the neighborhoods of the city through the inclusion of employment opportunities, political participation opportunities, and access to a close-in agricultural ring (located around the perimeter of the Garden City). The problem with the Garden City is that while it may have been viewed as a sustainable community it was not able to maintain the goal of social equity. The community was intended to house a range of social classes but due to financial pressures it was difficult to include lower income households. An example of the financial issues could be found in the increased land prices due to housing demand.

Similar to Clarence Perry and Ebenezer Howard, New Urbanists believe that communities should be walkable with a variety of destinations to which a resident can walk to and should incorporate a mix of housing to accommodate a range of incomes, lifestyles, and ages. Allowing for a mix of housing types that can accommodate a range of incomes and ages groups should allow a diversity of individuals the opportunity to choose walking as a mode of transportation however it appears that some New Urbanist communities may have problems similar to Howard's Garden City in that self-selection into the community can mean that housing prices become unaffordable to lower income households. Since the 1980s New Urbanism has been working to change the perception of pedestrian planning through the implementation of physical elements and design that allow walking to be an acceptable form of transportation.

3. Neighborhood design and walkability

One of the many ways that sustainability can be achieved is through the advancement of walkable neighborhoods which is a topic gaining in importance in both the planning and health fields as the activity of walking and the creation of walkable communities can have positive impacts on human health and the physical environment. While the planning and health fields may have initially emphasized walking for different reasons (for health professionals the emphasis was on health improvement and for planning it was for environmental improvement) the goals of the two research fields are ultimately complementary which is to improve quality of life.

To begin this discussion on walkability and its link to sustainable communities it is important to define the terminology used in related research. The three common terms used in planning and health related literature are walking, walkable, and walkability and while they may imply similar meanings there are differences between the terms walking and walkable/walkability. Walking refers to a physical activity done either for leisure or as a mode of transportation and the terms walkable and walkability are used to describe the degree to which the physical environment allows walking to take place. The portion of the physical environment often referred to when studying walkability is the space that is created by the streets, streetscapes, and building that are present in a specific location. A walkability audit is a tool for the assessment of the built environment to determine how it accommodates walking either by all of its residents or a specific target group such as the elderly.

In addition to the above terminology there are also two common assumptions about walking behavior; 1) that most people, when walking as a mode of transportation, will not walk farther then ¼ mile or five to ten minutes from their origination location, 2) when walking for transportation the route from the origination location to the destination should be as direct as possible. The two assumptions of route directness and length may not apply to both types of walking discussed in the planning and health related literature. As mentioned earlier walking may be for leisure purposes or as a mode of transportation. If done for leisure then route directness and length may not have as much influence on the decision of where and how long a person chooses to walk rather the physical environment may be the primary concern for leisure walkers. For those persons choosing to walk as a mode of transportation then route directness and length may be as important as the environment in which the walking takes place.

Beyond understanding what walkability there is another key element that is fundamental to the creation of neighborhoods that can positively influence a person's decision to walk. That key element is location-efficiency which refers to "...areas near transit, employment centers, or other essential services that allow families to reduce the number and extent of necessary car trips" (Haughey & Sherriff, 2010, p.2). The current suburban development pattern of exclusionary land-use is often far from being location-efficient particularly those developments that occur on the urban fringe. Access to public transportation may be non-existent and places to walk may be limited to other residential units or a nearby public open space when developments occur far from the city center. This lack of location-efficiency is demonstrated through the use of a GIS map (Figure 3) of Lincoln Nebraska showing the placement of grocery stores within a ½ mile radius of residential areas (orange parcels). The older pre-suburban sections of Lincoln which include the south central area of the city have grocery stores that are embedded within the residential areas. The newer section of the city, the bottom and lower right corner of the map, shows that grocery stores are now being located toward the edge of the residential developments. The changes in the location of grocery stores have made a difference in the amount of residential parcels that are within a

Fig. 3. Grocery store locations in relation to residential development in Lincoln, Nebraska, 2011. Created by Theresa Glanz.

walkable distance. It's important to note that while this map does not indicate there is a lack of grocery stores it does show that that there are gaps in the ability to be able to reach a store via walking. This may be particularly troublesome for elderly people that can no longer drive.

Beyond the terminology and assumptions mentioned above there is also concern within the research on walkable communities regarding the ability to identify the influence of the physical environment on human behavior. Two repeatedly mentioned research concerns are centered on physical determinism and self-selection and whether these are valid concerns. Physical determinism is the theory that the physical environment is responsible for the behaviors that occur within a given culture within a specific geographical location. Self-selection is the theory that an individual selects a location based on personal needs which may be financial, physical, and/or emotional. Both of these theories are mentioned in the research on walkable communities but their influence may be difficult to detect due to the inability to significantly separate observed behavior from personal preferences and from the presence or absence of environmental features.

Physical determinism is often a criticism of New Urbanist communities that make claim to creating environments that promote social community through the incorporation of walkable neighborhoods as there is not a consensus that design alone affects walkability. Self-selection is the ability of a person to select a specific location to live because it meets his/her needs. For example, a person may choose to live in a New Urbanist community because the physical environment is designed in a manner that allows a person to walk with greater freedom than in a conventional neighborhood. The importance of these research concerns are two-fold in that it is difficult to claim with absolute certainty the extent to which the physical environment influences human behavior while at the same time trying to determine the extent that self-selection plays in determining how a human behaves in a particular environment.

Additional criticism of New Urbanism focuses on whether the planners of these developments have actually created an environment that lives up to their claims of decreasing car-dependence, increasing pedestrian friendliness, and increasing the sense of community. Several studies have questioned whether new urbanism "...is too concerned with appearances...while ignoring social concerns..." (Southworth, 1997, p.28). The claim for decreasing car dependency seems to have some merit (Rohe, 2009), however, the claims of increasing socializing and sense of community seem to be harder to prove (du Toit et al., 2007; Hanna, 2009; Lawhon, 2009; Marrow-Jones et. al., 2004; Rohe, 2009; Southworth, 1997).

4. Assessing elements of neighborhood walkability

Measuring walkability is done either through the assessment of the physical environment (objectively) and/or through the gathering of personal perceptions (subjectively) of a specific location. The predominate method of gathering information for determining degree of walkability is done through the auditing of the physical environment which commonly includes features such as "building height, block length, and street and sidewalk width (Ewing, 2009)". These types of audits may also include observations regarding availability of street furniture, landscaping, physical condition of buildings, and cleanliness of area.

A second method of measuring for walkability is done through the gathering of perceptual information. This type of measurement examines a range of perceptual qualities held by the residents or users of the physical environment. The importance of completing a perceptual survey is that the researcher is able to gather information that is not readily available through the auditing of stationary objects. This allows the researcher to understand how perceptions affect the experience of walking and to gain an understanding of the relationship between perceptions and the physical environment (Ewing, 2009; Wood et al, 2010).

While there is two common ways of gathering information on walkability there are a number of physical criteria that can be or should be examined. First, research regarding walkable neighborhood design is more than just the presence of sidewalks and destinations to walk to it also includes macro and micro-scale features that affect the design of a neighborhood which in turn can affect the desire for physical activities, such as walking, by the residents (Alfonzo et al., 2008; Rodriguez et al, 2006). Macro-scale features include block length and number of intersections while the micro-scale features include street amenities, sidewalks, and conditions of the buildings in the neighborhood (Alfonzo et al, 2008). Together the macro-scale and micro-scale features can affect how the residents perceive the neighborhood environment (safety, pleasantness, accessibility etc) and these features may be found throughout the different neighborhood development patterns in the United States. While conventional suburban neighborhoods can be assessed easily for the above mentioned micro and macro features New Urbanism tries to incorporate physical features that go beyond those typical features by incorporating the following elements (Rohe, 2009):

- A street system that uses a grid or undulating design to maximize connectivity
- A mix of compatible land uses that includes housing, retail, and public facilities
- Single family homes set close to the street, with front porches, and garages set to the rear
- Pedestrian amenities and public open spaces

These features are incorporated into New Urbanism with the assumption that they (features) will encourage walking by the residents and socializing between neighbors. While New Urbanism, particularly at the neighborhood and street level, works to incorporate many of the design features that are thought to increase the desire for walking; within the literature on New Urbanism there is not a consensus that design alone affects walkability rather there is agreement that New Urbanism has created a lively debate about what makes a neighborhood/community sustainable, livable, and pedestrian-friendly (Morrow-Jones et al, 2004).

In addition to assessing the presence of sidewalks and building types available in a neighborhood Reid Ewing and Susan Handy (2009) have described five qualities that have particular importance when researching environmental perceptions; imageability, enclosure, human scale, transparency, and complexity. The quality of imageability refers to those features which help create an image of a particular place. This is highly personalized as individuals internalize perceptions differently; however, the social-cultural environment in which a person lives can create perceptual similarities when viewing an environment. Enclosure refers to the space created by the physical environment. Buildings, streets and sidewalks, and greenery such as trees can all provide definition of space. Ewing and Handy (2009) found that human scale was much more difficult to define than the previously two mentioned (imageability and enclosure) qualities. In part this is due to the differing opinions about what creates "human scale". Eventually Ewing and Handy (2009) list one of the definitions of human scale as "The size or proportion of a building element or space relative to the structural or functional dimensions of the human body. Used generally to refer to the building elements that are smaller in scale, more proportional to the human body, rather than monumental (or larger scale)."

The quality of transparency is the ability of the outdoor environment to project life within the indoor environment. It is a perceptual quality that allows a person to imagine what activities are taking place outside the direct line of sight. For instance, "courtyards, signs and buildings convey specifics uses (schools and churches) add to transparency" (Ewing and Handy, 2009, p.78).

Complexity is another quality that adds to the perception of the physical environment. This quality relates to the variation found within the environment and the ability of a person to internalize the information. Too little information creates boredom and too much creates information overload. Complexity is created by variations in the development pattern through varied setbacks, building orientations, and constructed buildings. Street furniture, signage, and the presence of and the activity of people all help to create complexity (Ewing and Handy, 2009).

In the book "Inclusive Urban Design: Streets for Life" by E. Burton and L. Mitchell (2006), there are six components discussed that promote walkability in a community. These components build on and expand the qualities mentioned by Ewing and Handy and they (components) are a mix of the physical as well as the perceptual. The following is a list of the components:

- Familiarity – refers to the extent that streets are understandable and recognizable.
- Legibility – refers to the ability of streets to help persons understand where they are at and which way they need to go.
- Safety – refers to the extent to which streets enable people to use, enjoy, and move around without fear of tripping or falling, being run-over or attacked. Safe streets have buildings facing onto them, separate bicycle lanes and wide, well-lit, plain, smooth surfaces.
- Comfort – refers to the extent to which streets enable people to visit places of their choice. Comfortable streets are calm, welcoming and pedestrian friendly.
- Accessibility – refers to the extent that streets enable a person to reach, enter, use and walk around places they need or wish to visit.
- Distinctiveness – refers to streets that give a clear image of where the person is, what are the streets uses and where they lead. (Overlapping similarities with imageability and complexity.)

In addition to the physical elements and perceptual qualities that affect the walkability of an area there are lesser discussed, but no less important, qualities that could be considered when studying walkability. Those qualities are the destination distance – how far does a person need to travel before they reach their intended destination; visit-ability – is the intended destination accessible by persons with varying physical abilities and weather – is the weather conducive to walking – year round, a portion of the year, or rarely.

5. Case Study: Neighborhood design and social interaction

In this section, we present a case study to show the relationship of walkable neighborhood design and social interactions. Sustainability of a neighborhood clearly contains the collective attributes of social interactions among residents. As shown (Cuthill, 2009; Dempsey et.al, 2009; Lehtonen, 2004), social sustainability is an important dimension of 'sustainable development', and is closely linked with environmental and economic sustainability.

In the Fall of 2010 a survey was undertaken in Lincoln, Nebraska to answer questions about the relationship between the specific variables of social interaction and walkability when the development design differed. Specifically the study sought to answer questions regarding 1) the amount of social interaction that occurs in two different types of neighborhoods, 2) whether walking by the residents occurs more frequently when the neighborhood design is based on New Urbanism principles, and 3) if a relationship between social interaction, walking, and urban design can be detected. The neighborhoods chosen were located in Lincoln, Nebraska and were comparable in age of development and housing prices. Sixty-three surveys were mailed out to households in the Village Gardens and the Wilderness Hills neighborhoods and of those 44.4% were returned.

Two neighborhoods were chosen for this case study; Village Gardens and Wilderness Hills which are located along the southeastern and southwestern edges of Lincoln Nebraska. Village Gardens is considered a Traditional Neighborhood Development which began construction in 2006 is located on the site of a former nursery. In addition to housing, Village Gardens has several specialty shops and a hotel that has been constructed and are now open for business; these are located in the northwestern corner of the development. Presently, the homes in Village Gardens consist of single-family homes and townhomes. The promotional website does indicate that apartments will be built however these will be restricted to the area designated as the Village Center (business center). The mix of housing is intended as a way of integrating a mix of incomes and lifestyles as well as being able to accommodate the changing needs of different life stages (Village Gardens, n.d.).

The second neighborhood used in this study is Wilderness Hills, a conventional suburban neighborhood, which is located along the southwestern edge of Lincoln Nebraska and is situated on former crop land. Construction in this development began in 2007 with the original phase nearly built-out and with subsequent development phases in the construction phase. Commercial development has occurred in the northwest corner of Wilderness Hills. Presently there is a big box retailer, a bank, and several constructed but unoccupied shops. The majority of homes built in this area are single-family homes with a few town homes present. Several of the lots in the second phase of development that had been designated for town homes have since been converted to allow for the construction of patio homes (these are homes that do not have to meet the minimum square footage requirements established for the single family homes within the development).

Four types of research methods were used to gather information for the case study; literature review, surveys, a walkability audit, and field observations of the two neighborhoods (Glanz, 2011). For data on social interaction a written survey was mailed to households in the Village Gardens and the Wilderness Hills neighborhoods. The survey was divided into three sections which included questions regarding interaction with neighbors, frequency of walking in the neighborhood, neighborhood satisfaction, and demographics. Participants are not identified in the results, however, in order to know which neighborhood the survey came from an identifier number was used on each survey; SE1 meant the survey came from Village Gardens and SW3 meant the survey came for Wilderness Hills.

To obtain information on the walkability of each neighborhood a walkability audit was completed for several streets in each of the two neighborhoods as well as photographs were taken of the areas. The walkability audit instrument was developed and was used to create an inventory of items as they related to sidewalk availability, location of house from street, handicap accessibility from the street, and presence or absence of people. This audit focused on elements and conditions that were readily observable which have the potential to influence a person's decision to walk. The following is a list of the conditions and elements that the walkability audit focused on:

- Surface conditions of the paved walking surface
- Path obstructions that would interfere with the ability to walk on the paved surface referenced above
- Segment features that may add to or detract from a person's desire to walk such as bus stops, street trees, street lights, and on-street parking
- Presence of litter, graffiti, or deterioration present in the observed area (condition of surroundings)
- The type of litter and disorder that may be present
- Whether people were visible and/or active in the observed area
- What, if any, crossing aids exist for aiding in the crossing of streets in the observed area
- The types of buildings and land uses that were observable
- The walking/cycling environment of the street segment which includes observing whether there were neighborhood watch signs, if there are bicycle lanes present, density of street trees, visibility of items such as trash cans or benches, and the depth of the building setbacks from the sidewalk
- Rating the overall attractiveness of the street segment which ranged from not attractive to very attractive

The features and conditions mentioned above work together to create an environment that a person may or may not find attractive to walk in and more importantly these can create an environment that a person would not feel safe in which in turn may deter a person from being outside.

The survey results revealed that there is a not a huge difference in the amount of social interaction among residents even when the neighborhood design differs. However, it does need to be noted that while the survey indicated that the respondents from both neighborhoods were comparable in knowing the same number of people and the amount of socializing that occurred; the respondents from the Traditional Neighborhood Development were generally more satisfied with the number of the acquaintances and friends they had within the neighborhood. (Glanz, 2011)

Two other points revealed by the survey were that the respondents in the conventional neighborhood ranked it higher as a good place to raise children (72.2%) then the TND (50%) however at the same time the respondents in the TND were generally more satisfied (80%) with their safety from threat of crime then those in the conventional neighborhood (61.1%).

It could be speculated that a higher satisfaction rate with the safety from crime in the TND could be due to having a higher satisfaction rate with their social relationships. As far as the difference in viewing the neighborhoods as a good place to raise children it may be attributed to the residents of the TND being older and as the survey revealed there are no children in the households in the TND that responded to the survey.

In addition to having greater satisfaction with social relationships this research showed a definite increase in the amount of walking that occurs in the TND over the conventional suburban neighborhood. Of the respondents, 80% of the TND respondents walked seven days a week while 61% of the conventional neighborhood respondents walked one to two days a week. The survey also revealed that the residents within the TND (80%) seen a greater frequency of people walking daily in the neighborhood then the conventional neighborhood (55.6%). This does confirm what is revealed in other published articles about walking and TNDs (Rohe, 2009). While this research does show that walking occurs with greater frequency within a Traditional Neighborhood Development it does not show if the design of the neighborhood has influenced the decision to walk or if the respondents are inclined to walk more than the residents in the conventional suburban neighborhood.

The physical audit of the two neighborhoods has revealed differences in the connectivity of the existing sidewalks. The sidewalk connection within the Wilderness Hills neighborhood was more complete and there was greater handicapped accessibility from the street as driveways could be used to reach the sidewalks. At the same time the resident survey showed a higher amount of walking in Village Gardens while having a lower amount of sidewalk connectivity. This does raise the question of the importance of the existence of sidewalks in residential neighborhoods.

Both of the neighborhoods in this study have a business center connected to them. Retail is a primary business category of both centers but there are important differences. The Wilderness Hills business center has several completed but empty buildings and does have the appearance that it is set up for predominately retail use with limited room for other types of business. This type of business restriction may limit the number of residents from the adjoining neighborhood who would walk to it. Village Gardens does have retail as a predominate business use but as indicated by the signs posted in the undeveloped areas of the business center there is the potential to create businesses that may encourage residents of this neighborhood to walk to them. A few of the empty lots are designated as areas that could be restaurants and one is marked for a specialty grocery store. Village Gardens has even been designed so that the distinction between residential and business uses is not as clear as in a conventional suburban neighborhood. This is done by incorporating housing units into the business center through the use of apartments above some of the stores and by having an alley instead of a street separate some of the residential units from the businesses and parking. This area of Village Gardens is more reflective of what Jane Jacobs liked about the inner city urban life – a mix of uses that includes residential. By mixing residential and businesses Jacobs coined the term "eyes on the street" which to her helped create a sense of security and was a factor in creating a sense of community.

If the presence or absence of sidewalks cannot explain the difference in the amount of walking that occurs than other factors such as the social aspects of the neighborhood should be examined with greater detail. This study does indicate that social factors such as satisfaction of neighbor relationships and safety from threat of crime may help explain why walking occur more often in different types of neighborhoods. As noted in an article written by Alfonzo et al. (2008, p.31) "It is unlikely that the built environment affects decisions to walk…rather…the built environment may support decisions to walk through the accumulation of several discrete features that together create a particular character or quality (safety, pleasantness, etc.)."

6. Conclusion

The past twenty years have seen only a handful of federal policies designed to help communities increase walking and biking opportunities within the United States. The Intermodal Surface Transportation Efficiency Act (ISTEA), the Transportation Equity Act for the 21st Century (TEA-21) and the Safe, Accountable, Flexible Transportation Equity Act: A Legacy for Users (SAFETEA-LU) (Handy and McCann, 2011) are all policies designed to enhance alternate transportation design. These policies however do not require communities to develop walkable surfaces rather they are funding sources which communities can use to help create walkable surfaces.

There are also no policies at the federal or state levels of government that require planners and developers to create compact, mixed-use, pedestrian neighborhoods nor are there policies that require the redevelopment of inner city areas over expanding cities onto greenfields. Currently the primary sources of support for New Urbanist communities are planners and developers that are willing to push for a form of neighborhood development that has not been common for several decades.

In addition to the limited federal policies on walking and biking there is also the issue of the public's response to compact, mixed-use developments. Information on public opinion towards compact development is not as readily available as the literature that expresses planners' opinions of this type of development. If a development is to succeed especially when the design does not fit the norm of a sprawling subdivision then planners need to understand who is most attracted to these types of developments, why are they attracted to them, and who is actually living in these developments.

With a lack of policies that mandate compact, mixed-use, pedestrian friendly neighborhoods there are two tools that are especially important for planners – education and marketing. With fuel prices rising now may be an especially important time for planners to educate the public about New Urbanism principles and how these principles, if implemented, can positively impact a person's life particularly in health and financial matters. The second tool, marketing, is important in helping to sell the concept of compact, mixed-use neighborhoods to people who may be interested in these types of developments but may not know of their existence.

This does leave the question as to whether the incorporating of New Urbanism principles should be required by public policy or whether it should be left to the free market to determine its usage. Presently these principles should be left to the free market but through better education of the public these principles may be more readily accepted by the general public as a means of helping them achieve a healthy and a less car-dependent lifestyle.

7. References

Alfonzo, M., Marlon, G.B., Day, K., McMillan, T., & Anderson, C. (2008). The Relationship of Neighbourhood Built Environment Features and Adult Parents' Walking. *Journal of Urban Design*, Vol.13, No.1, pp.29-51

Bureau of Labor Statistics. (2010). *Consumer Expenditure Survey*, June 2011, Available from <http://www.bls.gov/cex/#tables>

Burton, E. & Mitchell, L. (2006). *Inclusive Urban Design: Streets for Life*. Elsevier Science & Technology Books

Cuthill, M. (2009). Strengthening the 'social' in sustainable development: Developing a conceptual framework for social sustainability in a rapid urban growth region in Australia. *Sustainable Development*. Early View. Online published. DOI: 10.1002/sd.397.

Daniels, T.J. (2009). A Trail Across Time. *Journal of American Planning Association*, Vol.75, No.2, pp.178-192

Dempsey, N., Bramley, G., Power, S. & Brown, C. (2009). The Social Dimension of Sustainable Development: Defining Urban Social Sustainability. *Sustainable Development*. Early View. Online published. DOI: 10.1002/sd.417.

du Toit, L., Cerin, E., Leslie, E., & Owen, N. (2007). Does Walking in the Neighbourhood Enhance Local Sociability? *Urban Studies*. Vol.44, No.9, pp.1677-1695

Ewing, R. & Handy, S. (2009). Measuring the Unmeasurable: Urban Design Qualities Related to Walkability. *Journal of Urban Design*. Vol.14, No.1, pp.65-84

Glanz, T. (2011). *Walkability, Social Interaction, and Neighborhood Design*. Master Thesis. University of Nebraska, Lincoln, NE.

Hanna, K. S., Dale, A. & Ling, C. (2009). Social Capital and Quality of Place: Reflections on Growth and Change in a Small Town. *Local Environment*. Vol.14, No.1, pp.31-44

Haughey, R. & Sherriff, R. (2010). Challenges and Policy Options for Creating and Preserving Affordable Housing Near Transit and in Other Location-Efficient Areas, In: *National Housing Conference + Center for Housing Policy*, June 2011, Available from <http://www.nhc.org/media/files/chp_affordablehousing_TOD_challenges andoptions1.pdf>

Lawhon, L. L. (2009). The Neighborhood Unit: Physical Design or Physical Determinism? *Journal of Planning History*. Vol.8, No.2, pp.111-132

Lehtonen, M. (2004). The environmental-social interface of sustainable development: capabilities, social capital, institutions. *Ecological Economies*. Vol.49, No.2. pp.199-214.

Levittown Historical Society. (n.d.). Levittown history. June 2011. Available from <http://www.levittownhistoricalsociety.org>

Morrow-Jones, H. A., Irwin, E. G., & Roe, B. (2004). Consumer Preference for Neotraditional Neighborhood Characteristics. *Housing Policy Debate*. Vol.15, No.1, pp.171-202

Rodriguez, D.A., Khattak, A.J., & Evenson, K.R (2006). Can New Urbanism Encourage Physical Activity? *Journal of the American Planning Association*, Vol.72, No.1, pp.43-54

Rohe, W. M. (2009). From Local to Global: One Hundred Years of Neighborhood Planning. *Journal of the American Planning Association,* Vol.75, No.2, pp.209-230

Southworth, M. (1997). Walkable Suburbs? An Evaluation of Neotraditional Communities at the Urban Edge. *Journal of the American Planning Association,* Vol.63, No.1, pp.28-44

U.S. Energy Information Administration, Petroleum & Other Liquids, June 2011, Available from <http://www.eia.gov/state/seds/index.cfm>

Village Gardens. (n.d.). Village Garden Master Site Plan and Neighborhood Zones. In *The Plan.* March 2011. Available from <http://www.vglincoln.com/the_plan.htm>

Wood, L., Frank, L. D., & Giles-Corti, B. (2010). Sense of Community and Its Relationship With Walking and Neighborhood Design. *Social Science & Medicine,* Vol.70, pp.1381-1390

4

European Policy for the Promotion of Inland Waterway Transport – A Case Study of the Danube River

Svetlana Dj. Mihic and Aleksandar Andrejevic
Faculty of Business and Law Studies, Novi Sad
Serbia

1. Introduction

Sustainable development, as a global development concept, represents a multi-dimensional phenomenon and it includes many different indicators of human activities. When trying to view such a large concept it is necessary to individualize, measure and follow the movement of those indicators that are considered the most important and the most influential from the point of view of sustainability of future development. In certain number of cases, big changes in values of some indicators do not have a significant influence. However, indicators that show the state of the field which is exploited and use of energy belong to the group of the most important indicators of sustainable development in general [1].

In all fields of human activity, certain forms and amounts of energy are used in different ways and with a different efficiency degree, which depends on a big number of diverse factors. Anyhow, theory and practice show to the fact that transport in general is absolutely the biggest energy consumer and contributes to pollution in biggest amount [2]. That being said, a lot more attention needs to be devoted to consideration and implementation of solutions that will lead us to the lower energy consumption and lower exploitation for transportation needs.

In traditional sense, we can talk about air, land and water transport. Water transport is considered ecologically most acceptable for several reasons. Above all, in order to carry out water transport, natural waterways (rivers, canals, seas and oceans) are used, with the use of some waterways whose purpose is to shorten the distance during a certain trip. Another point is that, while conducting a water transport, many modern high-capacity means of transport are used and they allow heavy load transport. Apart from all of this, these means of transport can use ecologically acceptable fuels, especially biodiesel and its blends. Water transport, if conducted properly, does not jeopardize environment too much, it does not create waste, it does not create much pollution and it does not harm the view of the landscape, which can entirely retain its characteristics. Lastly, it is important to say that the economists today completely agree on one thing – water transport is absolutely the cheapest way of transport nowadays.

Because of everything above-mentioned, in all European countries, as well as inside the European Union, the possibilities to exploit and to use these existing waterways are seriously considered. Namely, the analyses show that there is an extremely well-developed network of waterways in Europe which are only partially used.

It has been estimated that around 30,000 kilometers of rivers and canals are running through Europe. They are evenly distributed in all European countries. Besides, in Europe there are some canals which were build on purpose and which connect north and south, east and west, Europe, Asia and the rest of the world. European rivers provide homes to some of the biggest and most developed capitals and cities. These areas are also famous for being the most developed ones and the most inhabited ones. In spite of favorable natural conditions, in the EU countries, water transport covers only 5.6% of total land transport in those countries. In the most developed European countries (which belong to Rhine region) water transport is constantly decreasing. From 12% in 1970 to 7% in the year 2000. At the same time, the total transport increased for 18% in the period of 30 years.

In order to promote the total transport, European Commission passed a so-called White Paper " European Transport Policy for 2010: Time to decide" by which Europe declares its willingness to intensify river transport as an economic, efficient, reliable and ecologically acceptable way of transport [3]. Likewise, the Declaration of European Ministers of Transport signed in Rotterdam in September 2001 called upon Member and Accession States to implement Pan - European RIS by the year 2005 [4] .

The European Parliament resolution following the White Paper sided the creation of high performance, geographically-comprehensive information systems of inland waterways to be extremely important in this connection and asked the Commission to submit a proposal for harmonized technical provisions towards the implementation of River Information Services (RIS). In the session of the Transport Council of 9th October 2003, The Netherlands, supported by other Member States, welcomed the Commission's initiative to put forward a proposal for a Directive on River Information Services. Meanwhile, this resulted in a RIS Directive, which creates a European-wide harmonized framework for River Traffic Information Services in order to ensure compatibility and interoperability between current and new RIS systems and to achieve effective interaction between different information services on inland waterways of international importance [5].

The harmonized river information services (RIS) on inland waterways in the Community Directive was published in the official journal of the European Union on 30th of September 2005 and came into force on 20th of October 2005.

2. Regional, continental and global relevance of Danube River

Danube is, together with River Volga, the longest European river. The length of the river from its spring in Germany, to its mouth in the Black Sea is 2,850 km. Danube connects 10 European countries. Taking into consideration the strategic concept of Europe as a region with long term sustainability strategy, European Commission has started considering the important potential, ecologic and economic relevance of unexploited waterway transport, by which the biggest attention is paid to the most important European river- Danube. River Danube is a waterway which makes an integral part of Trans European Transport Network (TEN-T). Via canal network, Danube connects Rotterdam harbor in the Netherlands with the Black Sea, that is, with Russia on the east. Because of all of this, Danube is thought o be the most important river in Europe, if not in the world.

Via Danube-Rhine-Main Canal, the length of waterway has been extended to 3,500km, and in that way western and south-eastern Europe have been completely connected. Danube has a navigable length of 2,411 km out of which 1,156 km (or 48%) are border sections. The countries which belong to Danube waterway are: Germany, Austria, Slovakia, Hungary, Croatia, Serbia, Romania, Bulgaria, Moldavia and Russia (via Black Sea). Given the position of Europe as a continent and considering the Danube's flow direction, this river can be seen as the "gate of Europe", that is it can be its water connection to Russia, Asia, Africa and via Mediterranean Sea it can be connected to the rest of the world (Fig. 1).

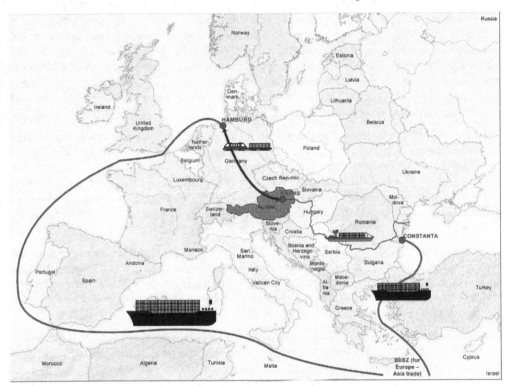

Fig. 1. Container Liner Services on Danube - Constanta as a gateway to Europe
Source: Graphics: Austrian Institute for Regional Studies and Spatial Planning

Apart from its global relevance, Danube's water flow is of vital importance for entire Europe and especially for the region and the countries through which it flows. Danube has multiple significance for the development of the region. Above all, Danube represents the cheapest and ecologically the most acceptable way of transport. This fact is especially important if taken into account that Danube flows through and connects some of the highly developed world countries which constantly increase their level of export and import services, which, of course, requires quality and efficient transport.

Besides, Danube embodies a sort of ecosystem with its own characteristics and regions and cities which it flows through have great historical and cultural importance and hence act as a backbone for sustainable tourism in this part of the world. Danube region and its countries are in part the most developed European countries and in part there are countries which are on the road to economic development. The prognoses point to the fact that the other Eastern European countries will grow economically and more intensely in the following decade. Therefore, contemporary Europe thinks of Danube as of the basis for development of this region, it sees it as a connection between European Union, Balkan countries in south and Russia, which is considered a region with special development potential.

All countries that are on Danube flow have marked a significant degree in economic development. Ninety million inhabitants live in the Danube region and they produce a Gross Domestic Product (GDP) of around 450 billion euro. Based on the predictions made by BMVIT and European Commission, the average growth rate degree in some countries of this region could happen in the period between 2010 and 2015 and it could be considered extremely favorable. Namely, all Danube region countries will show growth of GDP. This growth is predicted to be of 2.2% per inhabitant in Germany and even 4.9% per inhabitant which is expected in Croatia. This positive trend will impose the need for more intense and for a better organized river transport on Danube [6].

The European Union believes that Danube should represent the point of integration in this region and hence help and accelerate the progress of less developed countries, especially in Croatia, Serbia and Turkey, which are the only countries out of the EU at the moment. The countries through which Danube does not flow could also profit from the upgrading of water transport, in an indirect way at least. All of the Danube region is expected to show an extremely encouraging period of economic growth, and as a consequence it will have higher needs for transport.

There is a wide range of speeds that Danube's flow can take. Near the spring of the river the average speed is 6.5 km/h. Near Vienna the speed is 6 km/h. Afterwards the speed slightly decreases so that when entering Romania it is 4.6 km/h and on its way to the Black Sea the speed is 2.2 km/h. There are 78 bridges in total, whereas the biggest number of bridges is in bigger cities like Vienna and Budapest [7].

Regardless the fact that Danube does not flow through Russia, this country is extremely interested in Danube's waterway, especially given the fact that Russia borders with the Black Sea which represents the mouth of Danube river. Danube can be used for transport of natural resources and of products from Russia to the rest of the Europe and world, as well as vice versa. This fact shows how important is the need for a more precise long- term sustainable transport on this waterway.

3. Transport research on Danube River

Danube has been used for transport for a long time, but the precise data about transport on this river can be followed starting only from the year 1950, which was taken as the first year of the research. The final year of the research is 2005 or 2009 depending on the availability of the data. The data for the last year have been given in the form of an assumption, based on the movement on Danube in the last period, with a goal to get a clear and unique picture about character of transport on Danube in this period of 60 years, which was also the period of a strong economic development of Europe. The research covered the analysis of the following parameters [8]:

- Description and analysis of the existing fleet;
- The amount and analysis of heavy load transport;
- The amount and analysis of passenger transport:
- The influence these characteristics had on ecology.

The analysis of the data was made based on the research results, and suggestions have been given on how to improve the quality and quantity of traffic on Danube, to make its sustainable use possible by giving a significant contribution to sustainable economic and ecological development of entire west, central and south-east region of Europe.

3.1 Danube fleet

The development of Danube fleet has been monitored statistically since 1962. For the needs of this research the authors had at their disposal the data for the period since 1965, taking into account that some data are registered every 5 years, so that the data for year 2010 were not yet available. The research shows that the number and the power of the fleet has been increasing year after year. The basic changes in the characteristics of the fleet used on Danube in the period from 1962 to 2005 are the following:

- The number of the entire fleet grew from 3,142 vessels in 1962 to 4,529 vessels in 2005 which represents growth of 144%;
- The total heavy load which was transferred by the Danube fleet increased from 1,807,219 tons in 1962 to 4,385,986 tons in 2005 which represents growth of 242%;
- The biggest number of Danube fleet is used in Germany, Austria, Romania and Ukraine. The number of vessels in Serbia, Croatia and Moldavia is for 5% lower than the total number of vessels on Danube.

These mentioned growths point to the fact that Danube fleet transports more and more heavy load but not due to the bigger amount of vessels but because their capacity and operational power grow. From the point of view of sustainable transport on Danube this tendency can be considered extremely favorable, being that energetic efficiency could be one of the key solutions to problems of energy consumption today.

3.2 Heavy load transport

The volume of heavy load transport on Danube has been monitored since 1965 and it includes the analysis of the amount of transported goods and the analysis of traffic in harbors on Danube River. The analysis in this field gave extremely precious data when talking about the possibility of realization of water transport on this waterway. Namely, the objective of this research is promotion and stimulation of efficient and ecologically acceptable transport of bigger heavy loads by waterway. The amount of load has been monitored on three different bases: the load which entered the river's flow from the Black Sea, the load which used Danube to get to the Black Sea as well as the load transported between Danube harbors without reaching the Black Sea. By analyzing these data the authors concluded that the transport on Danube in the last 40 years recorded 3 characteristic periods, as shown in Fig. 2.

First and foremost, this diagram shows that the total amount of heavy load transported on Danube increased 4.8 times in comparison to the initial year, that is 1950. During this period,

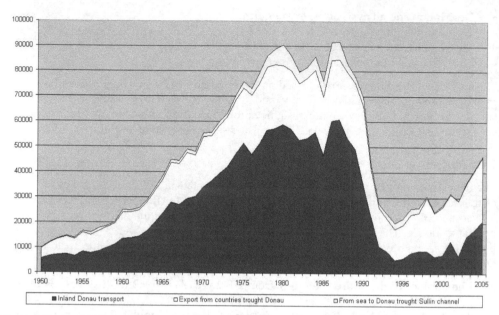

Fig. 2. The total of heavy load transported via Danube in the period from 1965 to 2005

the economic growth of some countries in this region directly affected the increase in transport as well. The growth registered each year reached its maximum in 1980 when a heavy load transport of 8 million tons was recorded. Such a high level of transported goods has been maintained until 1990. After that, there was a long period characterized by sudden decrease in the volume of transported goods, which is explained by the fact that transporters mostly used land transport. What brought to this decrease was also the worsening of economic and political situation and changes in countries of this region which belong to eastern Europe and in most cases they have accessed the process of political and economic transition. After the sudden decrease, the transport on Danube started recording some growth which was, however much slower so that in the last available year regarding the data(2005), the traffic equal to the traffic of 1970 was recorded. Because of these reasons, the promotion of Danube and sustainable transport development on it are of extreme importance.

The biggest volume of goods was recorded with the load entering the Danube through the Black sea. Somewhat less amount of goods refers to the goods transported between Danube harbors. The least amount of transported goods is represented by the goods which was imported from the countries of this region towards the Black Sea and further on. All this shows that in the past Danube, as well as now, has been used as a river by which different goods is imported in Danube region countries, and it is mostly referred to raw materials and unfinished products meant to be further processed. The least amount of goods is represented by raw materials or finished goods produced in the countries of this region and which are addressed to markets outside of Europe. It is very encouraging to see that transport among Danube harbors is well developed. The sustainable development of Danube traffic and this region as a whole insists on maximizing the use of Danube as means of road which can, whenever that is possible, substitute traditional land and railway traffic which represent much bigger a threat to the quality of the environment of this densely populated part of the world.

The biggest growth in the amount of transported goods was recorded in Romania, where transport increased 6.8 times respect to 1950. In the same period, the transport in Austria also increased, 5.7 times. The lowest growth was noted in Slovakia, 1.7 times.

Besides the general overview of transported goods in this monitored period, the traffic in harbors situated on Danube was also analyzed. The results of transport carried out in Danube harbors from 1965 to 2005 was given in Fig. 3.

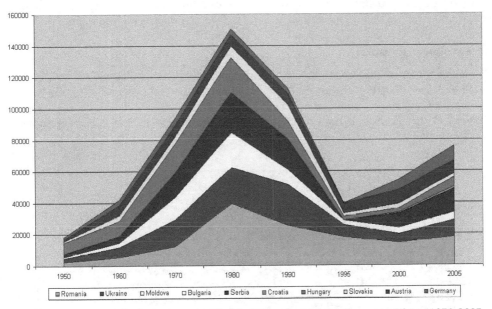

Fig. 3. The traffic in harbors on Danube divided by countries for the period from 1950-2005 (thousands of tons)

The traffic of goods in Danube harbors completely follows the tendency of transport on Danube in the monitored period. With the increase of transport of goods, the traffic in Danube harbors grew as well and it reached its maximum in 1980. After that the traffic in harbors suddenly decreases so that in 2000 it starts recording a new growth, with the expected positive tendency.

The biggest traffic was noted in harbors in Germany, Austria, Slovakia and Hungary, whereas the lowest values were recorded in Romania and Ukraine. Taking into account that Romania and Ukraine are as well the countries that have at their disposal 40% of entire Danube fleet, these results point to the fact that heavy load only travels through these countries but does not stop in their harbors. This situation opens the possibility for additional engagement of Ukrainian and Romanian harbors as a place for loading the goods produced in these countries. For the moment Danube in Romania and Ukraine has only a transition character. It is encouraging to know that traffic in Danube harbors in Serbia is increasing, even though it is a country with the least developed fleet in comparison to the other countries from this monitored sample.

The goods transported via Danube is very diverse. Mostly there are iron ore (25.6%), then processed and unprocessed metals (22.7%), coal (9.1%), oil and oil derivatives (8.5%), cement (7.5%), grain goods (6%), processed metals of metal industry (5.4%), natural resources: wood (4.3%), colored metals ores (3%), finished metal products (2.7%) and the least transported goods is agricultural goods like fodder (1.6%). These data show that Danube is not enough used for transport of agricultural products for no reason at all, especially given the fact that Danube flows through almost exclusively agricultural regions and where the production of food represents the basis for export of food from the countries of Danube region, above all for Hungary, Serbia, Croatia, Bulgaria and Romania.

3.3 Passenger transport

Passenger transport on Danube has been statistically followed from 1964 and the updated data have been available for the year 2005. Based on the general insight and for the needs of planning of sustainable transport on Danube, the analysis has been made regarding the changes in the number of transported passengers from 1964 to 2005 which is shown in Fig. 4.

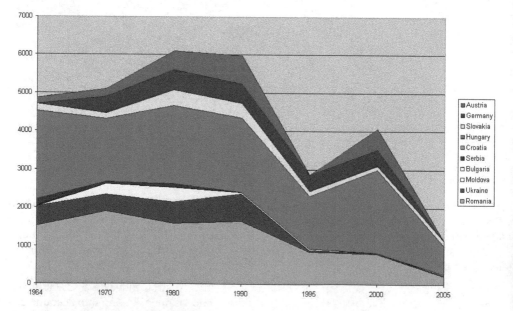

Fig. 4. Passenger transport on Danube from 1964 to 2005 (expressed in thousands)

Unlike the heavy load transport, where the constant growth was noted, passenger transport on Danube shows somewhat different characteristics. The biggest number of passengers was transported in the period between 1980 and 1990 (which coincides with the volume of the transported heavy load). Also, the biggest number of passengers was transported in Austria, Germany, Slovakia and Hungary. After that period, the passenger transport suddenly decreased and it reached its lowest point in 1995 which can be explained by the use of other, faster means of transport as well as by the wars and unstable political situation on the Balkans, given that general safety was very low and some of the bridges on Danube completely destroyed. After this period the number of passengers started growing rather insignificantly, for later it would start decreasing once again. The last year's data say that 1.2 million passengers were transported in 2005.

By monitoring countries, the biggest fall in number of passengers was noted in countries which at some point in past had the biggest number of passengers on Danube (Austria, Germany, Slovakia) which means that the promotion of water transport has to be adjusted to the requirements of the region. Also, based on this diagram it is easily noticeable that the number of passengers in Serbia is constantly low and it is especially surprising to know that there is a really small number of passengers in Romania, taking into consideration that Romania has big fleet. The fleet is primarily directed at heavy load transport, which means that planning the sustainable waterway transport must consider the investment projects to buy the fleet for transport of people.

4. Ecology and transport on Danube

Even though transport on Danube continues to show low intensity throughout the last 20 years the quality of the water has been harmed, being that there is a lot of waste water from many nearby industrial areas but also from cities located near the river. Regardless the strict laws which regulate the purification of water before letting it out, Danube has been significantly polluted. Given the impossibility to detect the sources of pollution, for the needs of the research the quality of the water has been decided not only based on the fact if it was pollution created by transport or in some other way.

Danube water in its upper waterway belongs to the II category (there is a moderate pollution in the river and good oxygen supply, there is a diversity of species and an abundance of algae, insect larvae, snails, entomostracans, aquatic plants which can cover wide areas and there is a lot of fish diversity). This type of water quality can be considered satisfactory, being that Danube in its upper waterway flows through Germany and Austria which are the countries that have reached very high level of development. Certainly, adequate laws in these countries (as well as in the EU in general) contribute a lot to these relatively good parameters of quality of Danube water in this area.

After going through Vienna, the capital of Austria, Danube water loses on quality and it changes to the III category (higher organic pollution; rather low oxygen content which affects fish; local sludge deposition; frequent and mass occurrence of sewage bacteria and ciliates; sponges, leeches and isopods can be often found; scarce aquatic vegetation). At some points going downstream from Vienna, the quality decreases even more and it become the IV category water (restricted living conditions for higher life forms; extremely high organic pollution; highly infected by organic, oxygen-depleting wastewater; total oxygen depletion is often caused by bacteria, flagellates and ciliates; suspended wastewater constituents cause turbidity or haziness; dense population of larvae and oligochaeta.)

Going through Hungary and reaching Belgrade, the capital of Serbia, the Danube water situation changes once again. It becomes the V category water and going downstream all the way to the Black Sea, Danube water is extremely polluted and at certain points it is even considered to be of VI or VII category [9]. Such a low water quality in this area is not so low only due to the highly developed industry, given that these are countries with a lower development level, but the insufficient and inadequate laws regarding environment.

The traffic on Danube affects the quality of Danube water depending on the quality of the fleet and on the frequency of ecological incidents. Given the age of the Danube fleet, European Union started a whole range of initiatives whose goal is to upgrade the quality of Danube water in its entire flow. For that reason, The Danube River Protection Convention has been enforced. It was created in the framework of the ECE – Convention for protection of trans-boundary waters that was enforced in 1998 and it is the only legal document regulating the transport on Danube and it includes Danube's tributaries as well. The basic objectives of this Convention are:

• Sustainable development and use of Danube as water resource and natural wealth of common interest
• Control of pollution of Danube
• Preventing further pollution of Danube
• Ecologically and economically justified use of Danube
• International collaboration
• Promotion of water transport
• The use of ecological energy as power generating fuel

The importance of Danube waterway for the EU can be seen in the fact that at the moment there are over 400 different projects on Danube which are being prepared or are already in process and their total value is around 5 billion dollars [10] and this money is to be distributed in the following way:

• Municipal waste water collection and treatment plants: 3.57 billion US dollars
• Industrial waste water treatment: 0.81 billion US$
• Agricultural projects and use of land: 0.16 billion US$
• Rehabilitation of wetlands: 1.12 billion US$

Bearing in mind that Danube waterway is very important and taking into account how much money has been invested into these projects, their realization is conducted under the supervision of the EU and it is conducted in 3 phases. In the first phase, completed in the end of 2003, the definition of Danube basin and Danube region was conducted together with conducting the analysis and upgrading the institutional frame necessary for collaboration among Danube areas and countries. The second phase of this enormous project was finished until the end of 2006, and it covered the analysis of the quality of Danube water, definition of ecology and source of pollution, as well as ways to control and monitor. After this phase there was a possibility to define clear strategic objectives and ways to realize them. In majority of Danube countries the third phase of the project is taking place. It covers conduction of certain activities. Given the character and volume of planned measures, the end of the first phase will be finished in 2021, even though there are some indications that it could last until 2027. All these things point to the need of clear and precise planning in all fields regarding the use of Danube, especially regarding the planning of sustainable traffic.

Its intensification has been predicted, and this fact in uncontrolled conditions can lead towards highly increased ecologic pollution.

5. Improvement in use of Danube River waterway

5.1 European policy for the promotion of inland waterway transport

In accordance with regulations conducted so far regarding the development of Europe, which is supported by the results obtained after the analysis of 60 years of Danube waterway, it is clear why all European countries support the development to intensify transport on Danube, certainly by respecting the concept of sustainability of transport and development in general. Considering the fact that Danube's potential has not been used enough, the new concept was defined as an all-embracing basis for the promotion of inland waterway transport. The name of the project is PLATINA.

Realization of Danube waterway promotion is connected to a lot of different problems which, above all, refer to the different levels of economic and institutional development of the countries through which Danube flows. The need for sustainable use of one river is possible only if all the activities, whose primary objective is equalization in position of some countries, are performed. Only in this way will it be possible to see Danube as unique, undivided and safe waterway, from source to mouth of river. For that reason, PLATINA projects points out the following primary activities :

- Establish a knowledge network bringing together all relevant actors concerned to assist in the implementation of NAIADES in Europe (EC, Member States and third countries),
- Provision of technical expertise and support,
- Provision of organizational, infrastructural and financial support and
- Platform deals with areas that require non-legislative coordinative actions at the European level.

These activities will be adjusted to the needs of each country. In order to proceed with efficient implementation of Danube promotion project, it is first necessary that every country makes its own personal action plan, which has to cover the following parts:

- Improvement and maintenance of waterway infrastructure

Each country that uses Danube waterway shows different quality of river and transport infrastructure. That being said, it is necessary that all the countries be put in the state which will suit the requirements of contemporary river transport [11].

- Exploitation of Danube's waterway potential

The results of this and similar researches confirm the fact that Danube's waterway is not being sufficiently used. Besides, the economic predictions of intensity regarding the economic development of Danube region in the following decades shows that there will be higher growth in industrial development and that imposes the need for more intense transport. Because of this, every country needs to think of Danube's waterway as of one of the most dominant existing waterways in the future [12].

- Building capacity for waterway and navigation authorities as well as for related administrations

The development of administrative and institutional systems which provide support and which intensify transport on Danube.

- Implementation of River Information Services on Danube

To create and implement the unique RIS system in all the Danube region countries represents one of the most important conditions for integration of the region [13]. In this way, Danube's waterway will surpass international borders and it will represent the unique waterway whose users will be able to use independent information and logistic support, which is something that will contribute to upgrading of the quality of sustainable transport.

- Implementation of Transport Management

The implementation of contemporary transport management primarily refers to providing the project with monitoring, control and navigation systems by which it is possible to control the traffic, to prevent delays, to reduce the cost of fuel and to reduce the cost of fleet maintenance [14, 15]. Besides fleet control, contemporary or modern transport management provides some solutions to the synchronization of harbors, load and unload equipment, container services and it provides the solution for bridges on Danube's waterway.

- Modernization of the fleet

The results of the research show that there are unfavorable characteristics when it comes to vessels used for transport on Danube. Especially worrying indicators of quality of the fleet are noted in the lowest developed countries (Serbia, Moldavia). It is necessary to proceed with adequate adjustments of existing vessels and with the purchase of new vessels which according to their characteristics fit with modern day economic, transport and ecological requirements. At this point it needs to be said that all the countries in the region have good conditions for production of biodiesel [16], which is considered excellent when used in naval engines.

- Development and integration of ecological strategies and concepts for Danube River

Danube represents a unique ecosystem with lower characteristics in some of its parts. Besides, some parts of Danube river bank represent protected natural, cultural and historical places. Sustainable development of transport on Danube needs to provide preservation and improvement of ecological maintenance of these regions.

- Creation of an international traffic model for the Danube region

It is necessary to analyze the existing infrastructure that is directly or indirectly related to Danube's waterway. All forms of land transport that are connected to Danube (roads, railways) have to be considered a priority in terms of investments for maintenance and for extending capacities. Also, it is necessary to decide future transport corridors which will connect industrial and agricultural capacities to Danube. Special attention needs to be given to modernization of international transport corridors, borders and customs free zones.

5.2 Applications and projects at work across Europe

The realization of PLATINA project proceeds via wide range of special projects whose goal is to promote certain aspects of inland waterway transport. All the mentioned projects are ongoing and each of them affects the creation of sustainable transport on Danube, as well as on other waterways in Europe. Due to the canal connection of rivers Danube, Rhine and Main, these projects cover almost entire territory of central, northern and southern Europe. The developments of national stand-alone telematics services, which vary in functions, standards and architecture, brought challenges to the current service regime. Some of the existing applications are [17, 18]:

- **ARGO** (Advanced River Navigation), a German navigation system for inland waterway skippers. It provides data on the fairway conditions and actual water levels in real time.
- **BICS** (Barge Information and Communication System) is a voyage and cargo (especially dangerous cargo) reporting system used in The Netherlands, Germany and other countries. The main aim of BICS is to support the reporting duties of the skipper/fleet operator towards the authorities.
- **BIVAS** (Inland Navigation Intelligent Demand and Supply System) is an internet-based interactive freight transport virtual marketplace.
- **DoRIS** (Donau River Information Services) is an Austrian system that can automatically generate traffic information by means of AIS transponders. The tactical traffic image is currently being tested for use by waterway authorities and skippers. In 2005 the roll-out of the DoRIS systems was performed on the Austrian section of the Danube and the operational test started in the beginning of 2006. Furthermore a subsidy program for RIS equipment is being provided for accelerating the penetration of RIS on the user-side.
- **ELWIS**, a German Electronic Waterway Information System, which provides a series of (fairway) information services.
- **IBIS** (Informatisering Binnenscheepvaart), a Flanders centralized database system, allows administrations to deliver navigation licenses, locate ships within their territory and collect data on inland navigation.
- **GINA** (Gestion Informatisée de la Navigation), a reporting application for Wallonia dedicated to the invoicing of navigation fees and the generation of statistics.
- **IVS90**, a ship reporting system used by Dutch waterway authorities supporting lock planning, vessel traffic services, calamity abatement and statistics.
- **NIF** (Nautischer Informations-Funk), a German service to transmit messages related to water levels, high-water notifications, water level predictions, ice and mist messages, and police messages.
- **VNF2000**, a French information network used to invoice navigation tolls and to produce traffic statistics.
- **VTS's Rhine**, Vessel traffic management services are installed on the Rhine in two difficult stretches: in the gorge section reach around the Lorelei in Germany with narrow bends and strong currents, and on the meandering Lower Rhine in The Netherlands with heavy traffic These differing operational practices and facilities in Member States reflect the current incompatibility of information systems, standards, and installations. Legislative and technical support for harmonized information services at a pan-European level become more and more necessary to guarantee the efficiency and safety for cross-border navigation and logistics. This was one of the principal motivating factors in the development.

Planned infrastructure projects:

- Germany: Straubing – Vilshofen (Danube km 2,321 – km 2,249)
- Austria: Vienna – cross border section with Slovakia (Danube km 1,921 – km 1,873)
- Hungary: Palkovicovo – Mohacs (Danube km 1,810 – km 1,433)
- Bulgaria - Romania: Iron Gate II – Calarasi (Danube km 863 – km 375)
- Romania: Calarasi – Braila (Danube km 375 – km 1,75)

The number and volume of mentioned investment projects show that modernization of Danube water way represents one of the most important initiatives which lead towards economic and ecological improvement of the region, which is based on sustainability and long-term stability.

5.3 Development and implementation of RIS

River Information Services (RIS) are the harmonized information services that support traffic and transport management in inland navigation, including interfaces to other transport modes. RIS do not deal principally with internal commercial activities between companies, but is available for interfacing with commercial processes. RIS streamline information exchange between public and private parties participating in inland waterborne transport. The information is shared on the basis of information and communication standards [19]. The information is used in different applications and systems for enhanced traffic or transport processes.

Modern logistics management requires extensive information exchange between partners in supply chains. Implementation of communications and information technologies in organizational and operational processes is a crucial prerequisite to increase operational efficiency and safety in today's market. RIS facilitate the inland waterway transport organization and management. Through effective information exchange, transport operations (such as trip schedules and terminal/lock operation plans) could easily be optimized, providing advantages for inland navigation and enabling it to be integrated into the intermodal logistic chains [20].

The degree of RIS system development in some Danube countries is different [21] but it is also conditioned by the influence of a big number of complex factors. The influence of the RIS development degree on the development of sustainable water transport is extremely intense, hence a short overview of its development in some countries is shown in further text.

Germany is one of the most developed world countries and it has a fully developed and implemented RIS system. The situation is similar in Austria as well. In this country there is a separate RIS centre which coordinates the functionality of the system in the entire territory of the country. The system of water level information, skipping service as well as availability of tactical traffic information all function. Electronic reporting system is still in development. The situation is similar in neighboring countries, Slovakia and Hungary [22, 23].

Somewhat more unfavorable situation regarding this issue is noted in Moldavia. Instead of fully developed RIS system, in Moldavia there is a tactical information transport center. There is no skipping service. In Romania all VTS centers are functional throughout the entire Danube waterway and there is also the skipping service. The full implementation of RIS system is in process and it is conducted with the help of the adequate EU institutions. In Bulgaria the situation is almost the same with the difference that in Bulgaria there is no skipping service.

Regardless of the fact that Croatia does not belong to EU, this country has a quality and well implemented RIS system. Water level and tactical information services are provided and there is also a skipping service. Electronic reporting system is in the process of development. The most unfavorable situation is in Serbia. In this country there is only one center that can provide with traffic information whereas the full implementation of RIS system still in process and it is supervised and helped by the EU.

5.4 The promotion of Danube waterway

The whole EU strategy is based on promotion and realization of conditions necessary for development of countries in this part of the world as well as all the activities conducted in them. Because of this, the special attention needs to be given to changes in traffic field which at the moment represents the biggest pollution threat for the environment. Based on the dimension of the problem and complexity of traffic in Europe, an all-inclusive sustainable development strategy, which concerns, apart from the other things, intensifying of water traffic in European waterways. These objectives are directed towards sustainable development of traffic and they include complex activities which have been going on for more than 20 years and therefore require special intensive and efficient promotion [24].

In Europe there are many organizations which deal with promotion of Danube waterway and they have the same objective to make Danube an intensive and ecologically more acceptable waterway in the future. If this balance of sustainability is reached, the stability in quality of Danube and river bank area, as well as improvement of quality of life of this area's inhabitants is provided. Given the importance of this problem and considering the number and diversity of Danube region countries, many governmental and non – governmental organizations follow and support the promotion of Danube waterway. Some of the most important are:

- Danube Commission with headquarters in Budapest, capital of Hungary, is at the moment the most important international organization which deals with regulation, promotion and implementation of sustainable transport on Danube. This organization consists of all the countries that can be found on Danube's waterway, including Russia, which is connected to Danube via Black Sea. The main source of finances of this commission is the fund of all country members.
- UN Economic Commission for Europe, Inland Transport Committee, the organization with headquarters in Geneva, Switzerland. The main goal of this organization is the technical development on inland transport as well as the development of laws and regulations in this area. This organization is financed by the UN.
- European Conference of Ministers of Transport gathers all the transport ministers of all European countries whose objective is the implementation of all sustainable and ecologically acceptable forms of transport on the territory of entire Europe. The sustainable development of traffic on Danube is one of the priorities of this organization whose headquarters are in Paris and is financed by European countries.

- Transport Infrastructure Needs Assessment: Corridor 7 is the association dealing with the development of transport infrastructure in inland waterways and it is financed by the EU Commission for the development of traffic. The headquarters of this organization are in Vienna, capital of Austria.
- Southeast European Cooperative Initiative: Danube Transport Working Group is the organization of Danube region countries with headquarters in Vienna. This organization deals with the development of traffic on Danube and it is financed by the EU member countries and by the USA.
- Stability pact for South Eastern Europe is a special organization in the EU, founded with the objective to support and provide stable development of countries in this region which were over the past years exposed to a number of negative influences. The headquarters of this organization are in Brussels, Belgium.
- Joint Austrian/Romanian initiative; Danube Co-operation Process is an association which stimulates the cooperation among all the countries which are situated on Danube waterway and its goal is to decrease border issues and limitations. It is financed by Riparian states, Stability Pact and European Commission.
- Via Donau is founded and financed by Austria, with headquarters in Vienna. The main goal of this association is improvement in quality and long-term sustainability of river traffic and implementation of RIS system on entire Danube waterway.

This is an overview of the most important organizations which deal with promotion of Danube waterway from different aspects. Their number, importance and financial sources speak about the relevance that sustainable development of Danube waterway has in all the countries of Danube region and in Europe in general.

6. Conclusion

The development of inland waterway transport is one of the long-term priorities of sustainable development of EU. Danube and Rhine- Main Canal as well as Danube's tributaries represent the most important waterway in Europe, therefore the biggest efforts of EU are put in analysis and improvement of sustainable water transport on Danube waterway.

By analyzing the available data which refer to the use of Danube waterway in period starting from 1950, it was stated that Danube waterway was used for the needs of traffic with different intensity. By analyzing the amount of freight transport, the number of passengers and heavy load in Danube harbors it is confirmed that Danube was intensely used as a waterway up until 1990 and after that its popularity decreased. After the year 2000, with the beginning of all-inclusive initiative of EU on promotion of inland water transport, some growth in use of Danube was noticed but that growth cannot be considered sufficient neither acceptable.

Therefore in Danube region countries, as well as in EU as a whole, many projects have been developed and there are many ongoing activities which were explained in this research. The objective of all these things is to promote and improve sustainable water transport on Danube. These activities mostly refer to modernization of Danube fleet (with emphasis on use of biodiesel as a power generating fuel), construction of adequate infrastructure, development and implementation of a unique information system on Danube as well as many promotional activities. EU predicts that the first phase of the activities, which deals with improvement of sustainable water transport on Danube, will be over in 2027.

7. References

[1] Mirjana Golusin, Olja Munitlak Ivanovic, *Definition, characteristics and state of indicators of sustainable development in countries of Southeastern Europe*, Agriculture, Ecosystems and Environment, vo. 130, issues 1-2,(2009), 67-74.

[2] Olja Munitlak Ivanovic, Mirjana Golusin, Sinisa Dodic, Jelena Dodic, *Perspectives of sustainable development in countries of Southeastern Europe*, Renewable and sustainable energy review, Vol. 13, Issue 8, (2009),pg 2179-2200.

[3] *White Paper "European Policy for 2020: Time to decide"*, European Commission *"Declaration of Rotterdam"*, Pan European Conference on Inland Waterway Transportation, September (2001) 58-62.

[4] DoRIS *"Donau River Information Services"*, http://www.doris.bmvit.gv.at/

[5] The strategic importance of the Danube for a sustainable development of the region, Workshop "Cross-programme ETC Danube projects", Austria, 2009.

[6] *Danube waterway – European key axis*, OSCE, Economic and Environmental forum, Austria, 2008.

[7] *Statistic ouverage 1950-2005*, Donau Commision

[8] Jürg Bloesch, The International Association for Danube Research (IAD)—portrait of a transboundary scientific NGO, Environmental Science and Pollution Research, Springer, (2009), 23-29.

[9] ICPDR / International Commission for the Protection of the Danube River

[10] PINE *"Prospects of Inland Navigation within the enlarged Europe"*, Feb 2003-Feb 2004, DG TREN project

[11] Reinhard Pfliegl via Donau - Danube Transport Development Agency Donau-City-Strasse 1, A-1220 Vienna, Austria

[12] *"Guidelines and Recommendations of River Information Services"*, PIANC, 2002.

[13] THEMIS *"Thematic Network in the Optimal Management of Intermodal Transportation Service"*, April 2000-April (2004), 5th Framework Programme thematic network, http://hermes.civil.auth.gr/themis/project.htm

[14] COST 326 *"Electronic Chart for Navigation"* - European cooperation in the fields of scientific and technical research, 1994-1997, http://www.cordis.lu/cost-transport/

[15] Doc Dr Svetlana Mihić Mr Saša Raletić, Kaizen kao način upravljanja performansama, XIV Medjunarodni naucni skup SM 2009 – Ekonomski fakultet Subotica, 21.05-22.05 Maj (2009) Palić – Subotica Zbornik spg.55 ISBN 867233224-5 COBISS.SR-ID 239684103

[16] PINE *"Prospects of inland Navigation in Enlarged Europe, SWP1.3 'Information and Communication systems'"*, Feb 2003-Feb (2004), DG TREN study

[17] EMBARC *"European Maritime study for Baseline and Advanced Regional and Coastal traffic management"*, Dec 2001-Nov 2004, 5th Framework Programme research project, http://www.euro-embarc.com/

[18] INDRIS "Inland Navigation Demonstrator for River Information Services", Jan 1998-June 2000, 4th Framework Programme research project

[19] ALSO DANUBE *"Advanced Logistic Solutions for the Danube Waterway"*May 2000- May 2003, 5th Framework Programme research project http://www.alsodanube.at

[20] COMPRIS "*Consortium Operational Management Platform for River Information Services*",
     Sept. 2002 – August 2005, 5th Framework Programme research project
     http://www.euro-compris.org
[21] BICS "*Barge Information and Communication System*", http://www.bics.nl
[22] D4D "*Data warehouse for the Danube waterway*",May 2001-May 2005, Interreg III B project,
     http://www.d4d.info
[23] SPIN-TN,"*Strategies to Promote Inland Navigation*", June 2000-June 2005, 5th Framework
     Programme thematic network, http://www.spin-network.org
[24] Doc. Dr Svetlana Mihić, Asis. Saša Raletić, Use of open inovation as possible marketing
     strategy in logistic, International Conference on Industrial Logistics " Logistic and
     Sustainability" March 8-th to 11-th, (2010) ICIL 2010 Rio de Janeiro Brazil ISSN:
     2177-0514 Proceedings page no.191

Part 2

Sustainable Tourism

5

Sustainable Tourism of Destination, Imperative Triangle Among: Competitiveness, Effective Management and Proper Financing

Mirela Mazilu

University of Craiova, Department of Geography
Romania

1. Introduction

Destination, or the terminus of tourists' holidays, is a complex element linking geography (with all the resources – anthropic and natural – made available to tourists) and tourism (with all the activities that they can carry out and services that they can consume during their stay). The complexity of the destination is that it represents a product and more products at the same time. Services forming the tourist product / products offered in a destination, and that must be differentiated from those offered by competitors, are brand "formative".

A tourist destination can mean a country, a region within a country, a city, village or resort. Whatever type of destination, the marketing tasks are the same: creating a favorable image of the destination at to the target segments of visitors, the design of instruments to support and disseminate the image and, last but not least, promoting the destination image in areas of origin. These topics will be covered in this article, focusing on the elements that define a tourist destination, the features of a destination, but also the elements that we insist upon in defining a destination image and its competitiveness, which confers a durability surplus.

Tourist destinations with limited financial resources for marketing activities are facing serious difficulties in producing an impact on the tourist market. It is therefore vital for the destination study to adopt a marketing policy such as "want it, get it", thus directing efforts towards clearly defined targets and using the most effective marketing tools.

2. Defining elements of tourist destination (destination image, destination competitiveness as tourism product, destination value-creating elements, destination identity)

2.1 Destination image

Attitudes, perceptions and images have an important role in the decision to choose a tourist destination. The image is the sum of perceptions and beliefs that people have in relation to that destination (Stăncioiu et al., 2009). The image of a destination is not necessarily based on prior experience, i.e. a visit to that destination. All tourist destinations have a self-image and the marketer's interest is to clearly distinguish it from other destinations, by defining components, by searching for key elements (Mazilu, 2010a, 2010b, 2010c) to transform it over time into a destination of a sustainable magnitude.

The ability to identify and promote value is the main factor of competitiveness of the society on a long-term. It requires identifying accurately and realistically the areas where there are performance premises and boosting quality development in relation to these areas, to identify individuals who are valued and their areas of excellence followed by channeling towards an education and training in line with their natural inclinations as to harmonize their interests with those of the society.

In conclusion, the competitiveness of a destination (Cândea et al., 2009) is given by the rate at which it manages to exploit the valuable human heritage with which it is naturally equipped. This has two major components: mass education and training and predisposition channeling, creating the conditions for the individual to manifest in that area.

Although economically speaking the concept of competitiveness in tourism defined as "the ability to cope with competition in an effective and profitable manner on the tourist market" incorporates that used in the literature, tourism specific content makes necessary a complex and multidimensional approach of this concept. This is necessary considering some particular aspects of the tourist product.

Firstly, we must emphasize that the multitude of components involved in designing and marketing a tourist product have made the achievement of its competitiveness to be a complex process in whose ensuring contribute: both the competitiveness of destination / resort, and the one made at each type tourist business: direct provider of tourist services: transport, accommodation, food, recreation, treatment, or intermediary: tour operator, travel agency, etc.

To this is added the fact that, from the tourist's point of view, the product covers the complete experience of leaving home and up to the return (Stancioiu, 2003), being sufficient that weaknesses manifest in a single component for the overall level of competitiveness to be affected. In this context, too, the first step of destination brand is to investigate and analyze the market and to formulate strategic recommendations related to situational factors, purpose and objectives established (Morgan & Pritchard, 2004). Of the factors of "success" on the strategic orientation of a destination, we can mention, among others, adopting a strategic vision, identifying key competitors, awareness / recognition of international competition, prioritizing infrastructure improvements, including the tourism development plan in the national development plan, taking into account the attitude / attitudes, cultural values and lifestyle of residents and, then of non-residents on their (own) city, etc.

This latter factor requires the marketer to carefully and constantly study visitors (including residents) and potential tourists (non-residents) on the image seen by them through the prism of cultural / tourist heritage of the destination as "the easiest and most effective promotion is self-promotion [...] which includes the strengthening of civic consciousness and self-confidence" (Ashworth, 2001: p. 58-70).

Sustainable Tourism of Destination, Imperative Triangle Among: Competitiveness, Effective
Management and Proper Financing

07

The image of a place is mainly "represented by its cultural heritage" (MacKay & Fesenmaier, 2000: p. 417-23) and formed by what is called "the mind or knowledge space" (Go & Van Fenema, 2006: p. 64-78). The cultural identity of a place is represented by its material and immaterial cultural heritage and, since the concepts of "culture" and "identity" cannot be separated, just as the concepts of "society" and "culture" cannot be divided – results the concept of "sociocultural" (Sorokin, 1967), the sociocultural identity of a people constituting, in fact, "its psychology or soul" (adapted from Luca, 2007: pp. 159-171).

Cultural identity of a place is an important part of its identity and therefore in its essence it must not change! In this context, the marketer's main problem is that, while identity remains and must remain the same over time, its image usually suffers changes. Therefore, the projected image must be realistic (Govers & Go, 2009: p. 190). In some situations, however, the image of a destination can be changed by "forming agents", such as news and popular culture (Gartner, 2009: p. 191-215) and may give rise to what is called the "a priori" image, one which, stable for a longer period of time, can change / distort even its sociocultural identity.

In general, the image of a destination can be built on a broad set of consumer functional and psychological expectations, on basic or holistic attributes, on common or unique aspects (Govers & Go, 2009, p.191), or on a combination of them. *If that place is wanted to become a successful tourist destination,* no matter of the model of tourist development and the proposed timeframe, *destination marketing plays a central place, the starting point being represented exactly by the inventory /* perception mode of its *"tourist heritage", in the smallest details, marketing audit constituting in this case an integral part of urban marketing.* 1

Value creating elements - from the perspective of the supplier / provider, transformed throughout the consumption of "a tourist product, viewed from the perspective of the buyer, in elements determined for satisfaction, compose inter alia **the image of a destination**.

The image of a destination is the sum of information and impressions submitted to potential consumers about the population, infrastructure, climate, history, attractions, personal safety, etc." (Echtner and Brent, 1991) and is formed by perceptions and experiences.

Since the sense of sight is predominant in forming a positive image, the visual perception of a destination can be decomposed, for a deeper understanding of the attributes / variables of differentiation that form the atmosphere / ambience, in an *artistic image* and a *psychological image*. Thus, intangible factors such as good weather, nature / scene, accessibility, turned into tangible elements, that is pleasant environment, relaxing ambiance, infrastructure, can create an artistic image favorable to the tourist choosing the destination. Intangible elements such as local culture, the diversity of sports activities, restaurants, cafes, etc., which can increase the value of a destination, turned into tangible elements, namely historical sites, events (cultural events, festivals, etc.) form an environment in which there are "lots of things to do", and may create a favorable psychological image for the tourist returning to the destination. Transforming these attributes into benefits for the tourist, by which a destination can be differentiated, is held in a positioning strategy.

---

1 This paper is an extension of a research made at the level of the tourist destination Romania (for the 8 regions), whose results are presented in the journal „Economie teoretică şi aplicată", starting with volume 2 / 2011.

Besides the differentiating variables "product" and "image", for a tourist destination other variables are also used (e.g., staff, a variable which can increase or decrease the value of the tourist product offered)2.

The destination identity is "the principal means of identification, but also the source of associations made by the consumer, which represent the links between values and brand" (Lindstrom, 2009). In the case of tourist destination, the identity elements are those which are constituted most of the times in attractiveness elements (which add value and / or uniqueness to the destination) and, at the same time, in main motivation for choosing it. The main feature of the selected destinations is their involvement in promoting social, cultural and environmentally sustainable models. The winners of this award are emerging European destinations, little known in the 27 Member States and candidate countries. The EDEN project helps to spread in the Union applied sustainable practices in the chosen destinations and to turn them into successful tourist destinations.

This project is supported by the European Commission, which launched it in 2006 and still plays the central role of coordination. The Commission's tasks consist in encouraging dialogue between stakeholders, co-financing selection procedures, organizing the awards ceremony (in the first two years, in the European Tourism Forum) and coordinating a comprehensive communication campaign. We have also benefitted from this support, and here I am referring to the area in which I work as President of the Association for the Promotion of Mehedinți County, occupying based on the international selection the fourth place, at a tie score (94) with three of the 32 participants.

Our major gain is outrunning greatly renowned tourist destinations, traditional not only for Romanian tourism, such as: the seaside, Brașov area, the area of Bucovina monasteries, etc. As a direct benefit of this gain is the free promotion at the European level of the European tourist destination: Drobeta-Ponoare-Clisura Dunării in European tourism fairs and other events, in this regard also being visited by foreign journalists who have contributed to the promotion of the above mentioned destination in reviving tourism in the area. The tourist destination is also accompanied by a brief bilingual description and a presentation CD.

A particularly favorable impact on our entire area, considered as an "open door" for future funding, the area enjoying a heavy promotion also at the European Union level, becoming one of the main destinations of foreign tourists coming to Romania. We also had facilitated the participation to three tourism fairs abroad with a customized stand and the participation at the Tourism Fair of Romania in autumn 2008. (economie.hotnews.ro/stiri-eurofonduri-4437193-depresiunea-horezu-premiata-comisia-europeana-**destinatie-excele**...

We must also highlight that often a tourist destination overlaps or is near a local community: a city or rural settlement whose economic, social, cultural life influences more or less tourist activity, being in turn influenced by it.

---

2 Adapted after Stăncioiu A.-F., Pârgaru I., Teodorescu N., Talpaș J., Răducan D. (2008) - „Imaginea și identitatea - instrumente de poziționare în marketingul destinației", paper presented at the Scientific Communications Session „Cercetare interdisciplinară în turismul românesc în contextul integrării europene", by the National Institute for Research and Development in Tourism, Ighiu, 2008;

Sustainable Tourism of Destination, Imperative Triangle Among: Competitiveness, Effective
Management and Proper Financing

89

Fig. 1. Drobeta Turnu Severin-Clisura Dunării-European Destination of Excellence

2.2 The competitiveness of a destination as tourism product

Competitiveness in tourism should be treated in the new conditions of globalization of economic life by highlighting the crucial elements that can influence and can become competitive advantages for Romanian tourism. The competitiveness horizon in Romanian tourism is inextricably bound to the elements of strategy adopted by the government through the National Tourism Authority, local administrations, each economic agent, elements which must combine organically with the sustainable development objectives.

All these aspects highlight the large number of determinants that influence the competitiveness of the tourist product.

Therefore, out of theoretical considerations, tackling the competitiveness of the tourist destination and tourist business will be made distinctly, although between the two there is mutual interdependence and interconditioning. A tourist destination cannot be competitive in the absence of competitive tourist businesses, while a tourist business cannot be competitive in an unattractive tourist destination. The crucial elements in ensuring competitiveness of the tourist destination would be schematically represented according to those in figure 2.

The model includes six determiners: the attractions and the tourist resources that the respective destination has; the support factors and other resources; the situational conditions/determiners; qualifiers; amplifiers; the conditions of the demand; the policy; the planning and the development of the destination; the inventory/the management of the tourist destination.

The attractions and the tourist resources existent at the level of a tourist destination include: the geographic position, the natural and anthropic tourist resources, the organization of some events, the relaxation and animation activities, the tourist equipments, and the commercial network dedicated to the tourists.

The success of a tourist destination is determined by the way in which this one manages to guarantee and at the same time to ensure its visitors, through its entire offer, an experience that can equal or exceed the multiple alternative destinations.

Building a cult for quality in tourism is a difficult process that needs the professional qualification of the personnel and an ethic education for the change of mentalities.

In order to achieve this, an education and motivation program of the staff is necessary divided on groups of professions and especially for the managerial levels, differentiated for those who will directly take part in the creation of a proper quality climate within the team, as well as showing attention, the desire to satisfy the needs, to answer to these needs as well as possible. This means among others: to apply the quality management, to completely involve the personnel of the unit regardless of the job and the qualification, to implement systems of evaluation and rewards, to elaborate rules and to educate the personnel etc.

The success of a tourist company, the effect of its competitiveness (see picture 2) are determined by the process of attraction, winning, satisfaction of the clients' needs, and especially by gaining their loyalty, offering good quality services and products. Following this pattern, the company will record the expected profit, following the effects of cooperation for the achievement of the competitiveness of a tourist.

As a result, ensuring competitiveness of tourist products and services must be based on *quality management*, it being a way to ensure competitiveness and hence business market credibility.

Starting from such an approach, we defined and outlined two concepts that those involved in tourism industry must learn and apply to come to produce market competitive services such as: competitiveness management and marketing in tourism.

Competitiveness in tourism suggests safety, efficiency, quality, high productivity, adaptability, success, modern management, superior products, low costs. A firm's competitive strength lies in competitive advantages and in distinctive competencies they possess relative to other competitors.

The success and more of a travel agency is determined by the process of attracting, gaining, satisfying customer needs, and also by their loyalty, the key to them being the quality of services / products offered; only in this way the company will be able to obtain the expected profit.

The essential objective of quality management is to achieve efficiently and effectively at a maximum level of only those **products that: fully satisfy customer requirements, comply with the society requirements, comply with standards and specifications applied, take**

Sustainable Tourism of Destination, Imperative Triangle Among: Competitiveness, Effective
Management and Proper Financing

91

into account all aspects of consumer and environment protection, are offered to the customers at the price and time agreed with them.

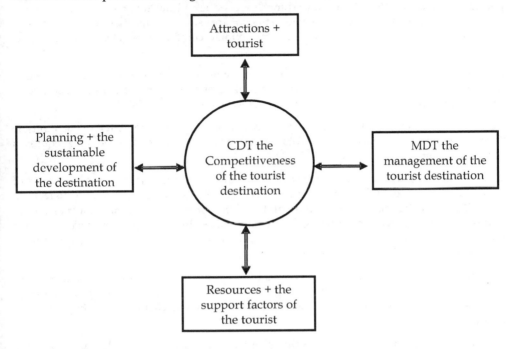

Fig. 2. Determiners in the competitiveness of the tourist destination

Introducing a quality system to benefit all parties involved: the country as a destination, entrepreneurs, consumers and intermediaries.

This includes ensuring a constant level of quality. Therefore quality implementation is achieved by a set of requirements called *standards*, and grouped by type, depending on the area in which they operate: performance standards, service standards, professional reference standards, standards including specifications, standards with operating procedures (operational).

Achieving quality involves not only developing standards and ensuring compliance with them, but quality performance must lead to meeting customer requirements and expectations of quality management.

In this context, the systematic analysis of tourist services / products quality and taking the necessary measures is a priority of utmost importance at the present stage.

Ensuring the competitiveness of tourist products is **the quintessence of the process of achieving competitiveness in tourism**. This is both the result of the competitiveness of providers directly involved in the production of tourism services included in the package, as well as of other determinants that have a bearing on the competitiveness of the tourist destination.

In this sense, marketers have defined the concept of **"product universe"** which summarizes very well these influences. Thus, the tourist product universe is represented by the sum of perceptions that the tourist has concerning the product: visual images: colors, ambience, geographical and physical environment, atmosphere, smells, musical sensations, human relations (with the personnel, other tourists, people), comfort level, etc. (Ambiehl et al., 2002).

A **product / service is competitive** when it has the ability to impose itself on a particular market, to sell in large quantities comparable perhaps with those of similar products or services produced and sold by competitors (Rondelli and Cojocaru, 2005).

As a result, the tourism product design is not confined to the combination of multiple variants of the two categories of elements: tourist resources and services, but also requires a **certain concept about the product**.

The ever-growing demands that tourists have to travel products have imposed a series of attributes that a tourist product must meet for it to be competitive: **satisfaction, accessibility, legitimacy, security and safety, authenticity, transparency, harmony with nature**, in their achievement bringing their contribution alike both competitiveness of each tourist service provider to the tourist destination and other competitiveness determinants of tourist destination (Cândea et al., 2009).

The differentiation firstly results from the value chain of the destination as tourist business (Fig. 3). As with any product, also in terms of tourist product, the value chain must be analyzed from two perspectives, namely: the business' value chain and the buyer's value chain.

From the business' perspective, the value chain includes the margin value and the value activities, each of which being a potential source of uniqueness. In turn, value activities (classified into primary and support activities) are divided into direct, indirect and quality assurance activities.

The evolution of the concept of quality, from quality assurance to total quality management, has required the use of working methods and procedures in all departments and at all the levels of the tourist product, with the establishment of indicators for all factors "creating satisfaction" (from quality ingredients for making food, respectively the quality of the linen cloth, up to the "quality of the destination", this also including service provider staff quality).

In other words, the concept of quality, approached from the perspective of the supplier / provider, found in all value-creating elements, is found in the form of the element of satisfaction for the buyer (e.g., degree of comfort - for accommodation, nutrition / taste value - for food, equipment degree, competence, etc. - for the destination as a whole). These, combined with the attractive natural resources, not necessarily unique, and constantly referring to the needs and desires of the market / markets that they wish to aim, lead to the differentiation of the destination with sustainable competitive advantages.

The new concepts on tourism development and a sustainable pattern of destination (Mazilu, 2010a, 2010b, 2010c) must take into account not only the varied and complex relationships between tourism and other regional economic and social phenomena, but must also refer to the phenomenon of tourism itself, as how it will shape in the future.

Sustainable Tourism of Destination, Imperative Triangle Among: Competitiveness, Effective
Management and Proper Financing

99

For Romania, this concept aims to take into account the evolution of social phenomena in the country, the financial crisis effects which under the terms of the market economy will generate the formation of new categories of potential tourists, of new motivations for leisure and consequently the occurrence and development of new tourist demands, new tourist destinations, which assume, under the global economic crisis, new tasks (see table 1)

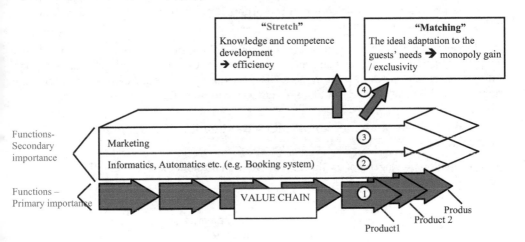

Fig. 3. Destination value chain as tourist business

Planning	Information	Establishment of tourist offer
• mission statement • organization	• information for guests • information for local people • support for journalists	• concern for guests • amusement for guests • coordination of tourism infrastructure • operation of amusement facilities • control and improvement of product quality
Marketing and communications	**Sales**	**Lobby**
• promotion • sales promotion • PR • brand management • market research	• information and reservation system • packages	• tourism awareness among the population • understanding of tourism at the level of political authorities • collaboration in "cooperative" organizations

Table 1. Tasks of a Sustainable Tourism Destination in the context of competitiveness

Romania's participation in the international tourism competition, on the continent and in the world, where there is very valuable tourist heritage at the level of European and world markets' requirements, remains a strong action issue of the government.

The systemic vision of the sustainable development strategy of Romanian tourism in the context of structural adjustment of the entire national economy enforces that tourism has become a prioritary economic sector in the organic interdependence with other branches and economic and social sectors.

The decisive element in the scientific and decision-making plan is to define a firm, realistic concept on capitalizing heritage and tourism sustainable development objectives. Romanian tourism alignment to these requirements is necessary both because of its characteristic mobility and because of the importance of this sector in Romania's economic recovery.

3. The destination management (DMO – Destination Management Organization)

The destination management is the coordinated management of all the elements which create a destination.

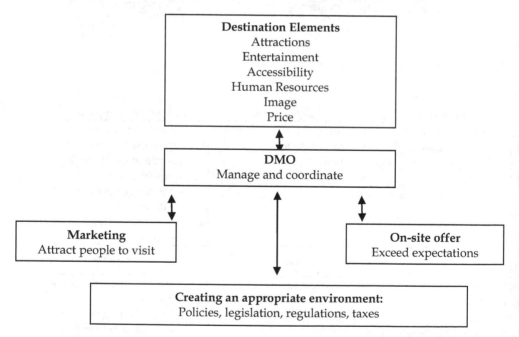

Fig. 4. The organization of the destination management – Stage I

The destination management approaches strategically these entities, sometime very separate, for a better result.

The coordinated management can help to avoid the doubling of the efforts regarding the promotion, the services offered to the visitors, the training, the support for business and the

Sustainable Tourism of Destination, Imperative Triangle Among: Competitiveness, Effective
Management and Proper Financing

105

identification and the management of the problems which have not been solved, complying with the following stages:

The organization of the destination management – Stage I

Synthesizing the models proposed in the studies, a possible model of competitiveness of the tourist destination is presented in picture 5.

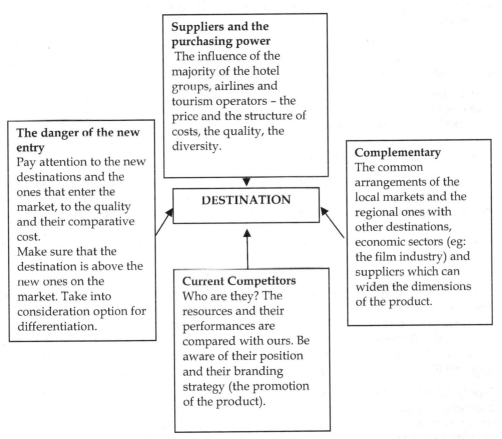

Suppliers and the purchasing power
The influence of the majority of the hotel groups, airlines and tourism operators – the price and the structure of costs, the quality, the diversity.

The danger of the new entry
Pay attention to the new destinations and the ones that enter the market, to the quality and their comparative cost.
Make sure that the destination is above the new ones on the market. Take into consideration option for differentiation.

DESTINATION

Complementary
The common arrangements of the local markets and the regional ones with other destinations, economic sectors (eg: the film industry) and suppliers which can widen the dimensions of the product.

Current Competitors
Who are they? The resources and their performances are compared with ours. Be aware of their position and their branding strategy (the promotion of the product).

Fig. 5. The effects of the Competitiveness of the Destination

The quality management is defined according to ISO 9000 as being "the assembly of activities of the general function of management which determines the policy in the field of quality, the objectives and the responsibilities, in order to implement them within the system of quality through specific means like: the planning of quality, the control of quality, ensuring the quality and the improvement of quality".

The essential objective of quality management is the achievement in conditions of maximum efficiency of those products which: entirely satisfy the client's requests, are in conformity with the requests of the society, with the standards and the applied specifications, take into account all the aspects regarding the protection of the consumer and of the environment, are offered to the client at the price and the deadline agreed together.

In 2010, Romania has recorded a success: it has climbed up 4 places in the top of the destinations preferred by the tourists from the EU in 2009.

Romania is currently on the 26th place in the ranking of destinations preferred by the tourists from the European Union, according to a recent report Euro-barometer, drawn up by the Gallup polling agency. Last year, Romania was the 30th destination in the preferences of the tourists from the European Union.

According to the statistics drawn up by Gallup in the 27 countries of the European Union, half of the 27000 respondents said that they would leave for holiday in their native country or in another EU country.

Criteria for choosing a destination depend on how it is perceived, namely the image that the consumer has established about it; at the same time, evaluation criteria of the quality of destination are also considered.

4. Sustainable destination management

It is essential for tourism development, especially through the efficient use of space and land planning, as well as through development control and the decisions to invest in infrastructure and services. Ensuring that new tourism development is, in scale and type, appropriate to the needs of the local community and environment, sustainable management can enhance on the long-term the economic performance and competitive positioning of a destination. It requires a supportive framework involving all stakeholders at regional and local levels, and an efficient structure to facilitate partnership and effective leadership.

A basic requirement for the existence and quality continuity of tourist destinations in Romania is to remain competitive. Actions to achieve this should be considered as part of the creation process of a lasting nature, which is one of the most important competitive advantages. Therefore, in order to ensure long-term competitiveness, viability and prosperity, tourist destinations should put more emphasis on full integration of concerns on sustainability in decision-making and in their management practices and tools.

Finally, to achieve a tangible progress, the demand from both leisure market and tourist businesses should send stronger and more consistent signals. Tourists need to be sensitized to be able to develop and strengthen their ability to make choices for sustainable development. Raising awareness with regard to sustainability and ethics can facilitate the emergence of tourist individual responsible attitudes and practices. The growing understanding of consumers in terms of durability could affect tourist destinations to show an interest in this direction and act accordingly, thus increasing their attractiveness, as in Romania's case.

4.1 Tourism development under sustainability

Tourism + sustainable development = sustainable tourism

At the beginning of this century and millennium, travel and tourism industry worldwide is the most dynamic sector of activity and, at the same time, the most important generator of

Sustainable Tourism of Destination, Imperative Triangle Among: Competitiveness, Effective
Management and Proper Financing

97

jobs. Economically, tourism is at the same time also serving as a source of recovery of national economies of those countries that have significant tourism resources and exploit them properly.

In this context, the main reasons causing the need for tourism development under sustainability, the following aspects are resulting (Mazilu, 2008a, 2008b, 2008c):

a. Tourist resources being practically inexhaustible, tourism is one of the economic sectors with real prospects of long-term development;
b. The complex exploitation and utilization of tourist resources accompanied by an effective external market promotion can be a source of increasing foreign exchange earnings, thereby contributing to the balance of external payments;
c. Tourism is a secure labor market and of redeployment of the dismissed one from other sectors heavily restructured;
d. Tourism is a means of promoting the image of a country, thus participating in the promotion of exports of goods and services on the world market, both implicitly and explicitly;
e. Tourism, through its multiplier effect, acts as a dynamic element of the global economic system, generating a specific request for goods and services that involve an increase in their production area, thus contributing to the diversification of the structure of national economy sectors.

One possible response to these challenges would be to apply the concept of **mosaic eco-development** (Mazilu, 2007a, 2007b, 2007c, 2007d), which proposes the implementation of sustainable development principles to smaller areas, following that they be gradually expanded so that, on the long term, to cover the entire national territory. In this view, environmental space should look in its ideal form like a chessboard, in which large pieces of agricultural land should merge with smaller areas allocated to industry, various categories of infrastructure and parks and natural reserves. This complex alternation arises from uneven spatial distribution of natural resources and the application of economic, social and environmental criteria. In this framework, ecology and bio-economy can provide original solutions to territory arrangement so that appropriate environmental areas are allocated to each branch, resulting in a territorially sectoral complementarily.

The resulting complementarity should be addressed not only functionally but also by the rational use of land, by the increasing of employment level and income, by the actual participation in inter-regional exchanges and integration into European structures and flows, by the complementarity with environmental restrictions (Constantin D.L., 2000).

Application of sustainable tourism development projects starts from the design and construction of the material and technical base in order to harmonize with the environment, local community or other sectors of the economy, continuing in the state of tourist activities course.

Seeking a sustainable development model appropriate to each territory, not universally valid and applicable in any territory, we observe analyzing the etymological game but with a lot of sense: Tourism + Sustainable Development = Sustainable Tourism, the insertion of constraints, similarities, differences, which requires an even closer analysis. It is known that each area has its history, its identity, its resources, according to which a form or other of tourism can develop, imperatively respecting its economic, social and environmental specificities.

In this perspective, the different players "employed" in such an approach are called to establish, in a first phase, a careful diagnosis of the territory, aiming even a decoding of the influences of sustainable development according to the opportunities and threats of each territory. A second phase is the drafting of a sustainable tourism development project of that territory continuously adapted to the local context, the project being "embraced" by as many people living in that territory possible. Last but not least, certain "clauses" of territorial development will have to be respected, progressively putting in place actions aimed at improving public offer, the local tourist product, removing the parasitizing of this sustainable action of other illogical ones (power games, interest games among stakeholders, etc.).

This article, open itself to major scientific reflections, has tried to explain the role and importance of the analytical approach of sustainable development in tourism, in the territory, having the function of restructuring it, rebuilding it and even redeveloping it into a better direction, because sustainable development respects the direction of history being directed towards the future.

There is no single model universally applicable for the sustainable development of tourist destinations (Mazilu, 2010a, 2010b, 2010c). In this perspective, the different players involved in this inseparable binomial: Sustainable Development - Tourism, are required to build in these territories **a specific tourist offer** to meet on the one hand individual or multiple demands, and on the other hand to meet local crisis (economic, political, social, etc.).

Tourism can contribute to sustainable development of territories because the territories themselves are part of an interactive, integrated and responsible relationship with the economic, social and natural environment on which they depend. Certainly, more remains to be done, because the unbalanced and destabilizing effects (see fig. 6) and the resistance to change, unfortunately, still persist, despite the massive involvement of the local community towards sustainable development.

In sustainable development, tourism plays a key role in contributing with a high percentage to Romania's economic revival and recovery. Raising tourist product from established values to those corresponding to quality standards and preferences (Mazilu et al., 2010) of foreign tourists involves initiating and promoting actions to include, on the one hand, the development of education and training processes of mindsets appropriate to the current type of development, and, on the other hand, increasing sustainable development in tourist reception regions.

4.2 Stabilizing (+) and de-stabilizing (-) effects of tourism on the sustainable development

In global world you must live globally. Or integration is impossible without learning the rules of world tourism, without also learning to follow the code of conduct. One is especially not allowed to ignore that the effect of tourism regards only the future. The present's sensations are sublimated, memories become past: a past that will determine future actions. Tourism is not only a school about others, but determines how we will live with others, how we will behave. Our world, one created by tourism producers is a global one, a single ethnic group: the human race in its specificity, item by item for diversity. With a past, a present and hopes. Any mistake can lead to incurable trauma (ARA)3. And this on a mass of people called tourists.

The concept of sustainable development is today environmental. On the other hand it forces us to ask ourselves of the social purpose of our acts and of the future of the planet, taking into account permanently the economic system affected. It, as sustainable development, appears to us as a pioneer to new reflections, new actions, thus dedicating a mindset, even of government, unusual, based on cooperation and negotiation with all stakeholders in the implementation of sustainability in the territory (see fig. 7).

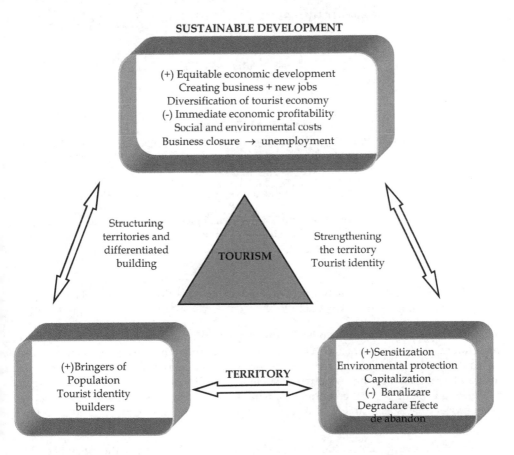

Fig. 6. Stabilizing (+) and destabilizing (-) effects of tourism on sustainable development (Author's adapting after Oliver Bessy – Sport, Loisir, Tourisme et développement durable des territoires, PUS, 2008, Cedex, p. 44)

---

3 ARA

On the other hand, sustainable development can be revealed to us as an alibi, even a utopia maintained voluntarily by political and economic players, anxious to legitimize and prove their economic logic but also to preserve the economic strength already acquired. We believe that to resolve this situation it would be enough to adapt socio-economic and ecological systems of this globalization "given": **Sustainable development**.

In fact, this ambivalence reflects perfectly the state of our society harassed back and forth between a dominant liberal model organized around "tourist market" (Cândea et al., 2009) but also to a welcome closeness, regulating in the environmental and human level.

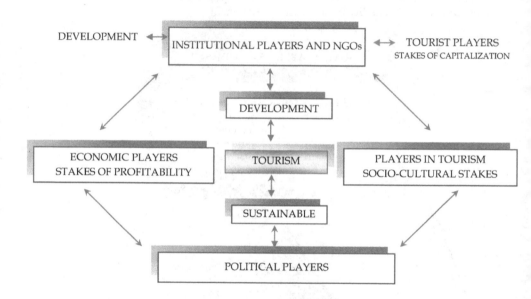

Fig. 7. Foreground stakes of the different local players in the action for sustainable development of tourism (Pârgaru et al., 2009)

Sustainability for tourism as for other industries has three independent aspects: economic, sociocultural and environmental. Sustainable development implies permanence, which means that sustainable tourism requires the optimal use of resources (including biological diversity), minimizing the negative economic, sociocultural and ecological impact, maximizing benefits of local communities, national economies and conservation of nature. As a consequence, sustainability also refers to the management structures needed to meet these goals.

The goal of achieving sustainable tourism should be subordinated to national and regional plans of economic and social development. Actions may cover for economic goals (income growth, diversification and integration of activities, control, development potentiation and zoning), social goals (poverty and income distribution inequality improvement, indigenous sociocultural heritage protection, participation and involvement of local communities) or environmental goals (protection of ecotourisms functions, conservation and sustainable use

Sustainable Tourism of Destination, Imperative Triangle Among: Competitiveness, Effective
Management and Proper Financing

101

of biodiversity). Some experts prefer to speak about sustainable tourism development rather than sustainable tourism, the first referring to all aspects of development, and the second to some aspects and components of tourism – such as long distance air transport that may easily not be sustainable under current technologies, even using the best practices.

We also wish to recall which are **the goals of sustainable tourism**, as they were mentioned by the WTO, in 2005 (acc. to *Making tourism more sustainable, A guide for policy makers*, UNWTO / UNEP, in the Sustainable Tourism Group report, "Action for more sustainable European tourism", published in February 2007 – Annex 2 of the document cited includes the **12 objectives** to be met by development of sustainable tourism activities in a tourism destination with a protected area title):

- **Economic viability**

To ensure viability and competitiveness of tourism destinations and enterprises so that they are able to continually prosper and provide long term benefits.

- **Local prosperity**

To maximize the contribution of tourism to the prosperity of the host destination, including the proportion of visitor spending which is due to locals.

- **Jobs quality**

To increase the number and quality of jobs created and supported by tourism locally, including salary, working conditions and their availability to all persons without discrimination of gender, race, disability or otherwise.

- **Social equity**

To reach a broad distribution of economic and social benefits in tourism to the recipient community, including opportunities, income and services growth of those less wealthy.

- **Visitor satisfaction**

To provide a safe, satisfactory and complete experience for visitors, which is available to all without discrimination of gender, race, disability, etc.

- **Local control**

Local communities to be involved and empowered in planning and decision making on the management and future development of tourism in their area, in consultation with all stakeholders.

- **Community welfare**

To maintain and enhance quality of life in local communities, including social structures and access to resources, attractions and life support systems, avoiding any form of social degradation or exploitation.

- **Cultural richness**

To respected and put forward historical heritage, authentic culture , local traditions and specificity of host communities.

- **Physical integrity**

To maintain and put forward quality of landscape, both of the urban and rural ones, and to avoid physical and visual degradation of the environment.

- **Biological diversity**

To support preservation of natural areas, habitats and wildlife and to minimize their damaging effects.

- **Resource efficiency**

To minimize the use of scarce resources and non-renewable resources in the development and operation of tourism facilities and services.

- **Environmental purity**

To minimize the effects of air, water and soil pollution and waste production by tourism enterprises and visitors

To be mentioned that, for such requirements, a responsible, judicial tourist is also imperative for their compliance, to observe them, to manage them: better said – a sustainable tourist.

However, sustainable tourists are people taking into account sustainable development pillars when defining the trip's tourist package – i.e. accommodation, transport and activities – and are respectful towards nature, culture, people and destinations.

The traveler must behave respectfully during holidays, as well as in the case of an event related to his profession or at a congress, both for residential tourism and one-day travel.

The stereotypical image of the sustainable traveler is often associated with backpackers. However, this picture is not correct and complete, having nothing to do with the travel manner (individual or group), travel organization (individual organization or organizing by specialized intermediate agents), type of accommodation, type of vehicles and even with the destination itself. The condition is to make conscious choices, taking into account the said principles of sustainability. However, these conscious choices must be visible through the nature of the destination.

To behave respectfully towards other people, culture and environment involves, for example, taking an ethical behavioral conduct, using facilities enjoyed by locals, taking into account local customs, adapting attire where necessary, refusing to buy certain souvenirs that could harm the place. It also means paying a fair price for services. Such behavior could, for example, inspire tour operators to include in their offer sustainable tourism products.

A sustainable traveler is aware that during his trip his conduct has effects in the destination country.

Social implications of a sustainable travel to avoid oversaturated tourist traffic are important especially in areas where cultures substantially differs from the external environment of the area visited.

Sustainable Tourism of Destination, Imperative Triangle Among: Competitiveness, Effective
Management and Proper Financing

103

- Proposals:

Countries with natural and human tourism potential similar in richness and variety with Romania – sometimes not so diversified and focused – have managed to make tourism industry an important factor, in some cases even the most important, of economic growth and general development.

To achieve this target it is necessary to act in the following ways:

- make the institutional reform and create the legal framework for decentralization of management;
- harmonize tourist legislation similar to the EU one;
- support the establishment of professional associations and other NGOs in tourism, open and organize the National Authority for Tourism partnership with them;
- raising the quality of tourist promotion actions by using PHARE funds, supplementing the insufficient budgetary funds (compared to European Union countries) for this area.

Tourist policy on the medium and long term will have to aim the following priority objectives:

- reduce taxation;
- continuing to maintain international tourism as an export activity;
- tax exemption on reinvested profit on a certain period;
- further improvement of the legal and institutional framework for the harmonization with the rules of the World Tourism Organization and the European Union;
- state involvement in financially sustaining investment in tourism, particularly those of public interest (infrastructure), as well as international and domestic tourist promotion;
- development of the specialist professional training system and professional reconversion for dismissed labor force in other economic sectors; setting up a network of tourist education establishments integrated in the European network of hotels and tourist education;
- correlation of programs and tourism development projects with regional development programs (transport, telecommunications, landscaping etc.).
- increased attention to polls – useful tool for hotel managers working to maintain and increase the quality of services rendered.
- imposing quality marks in order to increase competitiveness on the tourism market and tourism service quality recognition.

Sustainable tourism represents tourism that meets the needs of present tourist and host regions while protecting and enhancing opportunities for the future (http://www.ecotourism.org). On the other hand, sustainable tourism is a necessity to apply the principles of sustainable tourism development. The concept of sustainable tourism has become a key ingredient in the nation's tourism strategy (Connel et al, 2009, p.867).First, Sustainable tourism is the ability of a tourist destination to remain competitive (Mazilu et al., 2009), while maintaining environmental quality, despite all problems, to attract visitors for the first time, then following that loyalty to him, to remain unique in terms of cultural and be in a permanent equilibrium with the environment (Nistoreanu, 2009). It is based on the consideration that development must meet the present needs without jeopardizing those of future generations (Nistoreanu, 2005, p.42).

Sustainable development and its derivative sustainable tourism, although intuitively appealing, widely adopted by international organizations and many governments, and enshrined in legislation, are concepts that have been much criticized because of their lack of precision and because of the difficulties that have been experienced with their implementation (Tao & Wall, 2009, p.90).

Sustainable tourism develops the idea of meeting the needs of current tourists and tourism industry while protecting the environment and opportunities for the future needs to achieve the entire economic, social, aesthetic, etc., actors in tourism, while maintaining cultural integrity, environmental, biological diversity and all systems that support life (Stănciulescu, 2004, p.23). Sustainable development plans cover at least three (Ionescu, 2000, p.137):

- Economically, by increasing service and resource recovery;
- Environmentally, by recycling, avoiding environmental degradation, reduction of land removed from agricultural use;
- Social, by increasing employment, practicing traditional population to attract tourism, as measures of physical and political regeneration (Figure 8).

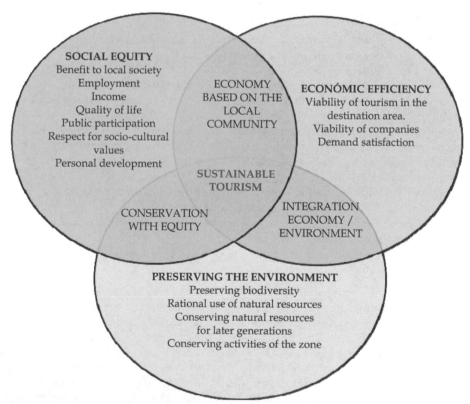

Fig. 8. Sustainable Tourism Model (Source: Sanagustín Fons M.V. Moseñe Fierro J.A., Gómez y Patiño M. – "Rural tourism: A sustainable alternative", 2011, p. 552)

Sustainable Tourism of Destination, Imperative Triangle Among: Competitiveness, Effective
Management and Proper Financing

105

Since **sustainable tourism is an end**, one must understand that any type of development including tourism development gives rise to certain changes in an area. However, these changes should be kept within acceptable limits, so that sustainability is achieved. Sustainable tourism can be best achieved by careful planning, by development and proper management of the tourism sector based on some principles (see Principles of sustainable tourism in World Tourism Organization, WTTC, The Earth Council – Agenda 21, 1995).

Thus, by 2050 the world population today estimated to over 6.5 billion people shall be between 7.7 billion and 11.2 billion. However, the latest average projection is of 9.37 billion people.

Population growth is the main pressure on the environment and it is inexorable.

Tourism industry is seen by its specificity as environmentally tied more than other industries, but its size and presence have created negative physical and social impacts on the environment.

Tourism, like any human activity, participates in degradation and environmental pollution, either by direct pressure of tourists on the landscape or on other tourist attractions, or by the misconception of capitalization of tourist points and attractions. The impact of tourism on the environment is determined by:

- uncontrolled tourist traffic, chaotic in areas or tourist objectives outside marked trails, leading to destruction of vegetation, flora and fauna;
- non-systematized development of settlements reaching up to an excessive urbanization of resorts (Baile Herculane, Baile Felix, Sovata, Bran, etc.), over-sizing of resorts in terms of reception and treatment capacity (Baile Felix, Sovata, Buziaş, Vatra Dorna, Sinaia, Poiana Brasov, etc.)
- changing the physical and chemical parameters of therapeutic resources (Vatra Dornei, Buziaş, Călimăneşti, Sovata, Ocna Sibiului, etc.) and treatment muds (Techirghiol, Negru and Ursu lakes in Sovata and Săcelu – Gorj);
- lack of rural wastewater treatment plants, ecologic landfills for waste and flood protections;
- strong influence on the aquatic environment in tourist resorts areas on the Romanian coast and the Danube Delta;
- total or partial damage of caves due to their arrangement for visiting, made without complying with scientific techniques of such works Muierii Cave (Parang Mountains), Ialomita Cave (Bucegi Mountains) and Peştera lui Ionel (Bihor Mountains);
- automobile tourism entering tourist resorts (Călimăneşti-Căciulata, Tusnad, Vatra Dornei, Sinaia, Busteni, Predeal, etc..).

Or, to avoid such "disasters", which happened in destination areas, I frankly think that we need tourism to integrate the natural, cultural and human environment and respect the fragile balance, characteristic to many tourist destinations.

Consequently, the need for new and professional tourism leadership emerged, to attract more the governments and the private and public sector partners based on **principles** of sustainable development, namely:

- environment has an intrinsic value that is especially large for tourism, that the next generation should benefit from, too;
- tourism must be considered a positive activity with several beneficiaries:
  - environment
  - local communities
  - visitors;
- the relation between environment and tourism can be developed so that environment supports tourist activity in the long run; in turns, tourism is "forced" not to cause due its performance the degradation of the environment;
- the development of the tourism activity must comply with the ecologic, social, economic, cultural features of the geographic space where it is performed;
- the purpose of the tourism development must always be to counterbalance the needs of tourists and the ones of destinations and of their hosts;
- the tourist industry, the governments, the authorities responsible for the environment protection and the international bodies should comply with these principles and work together to implement them.

Polyvalence is quite necessary for the environment and tourism specialists, especially for the ecotourism specialists, due to the fact that futurologists have been using lately the following syntagm: "too much tourism kills tourism", highlighting the fact that there are "limits" in the process of tourism.

If we do not want tourism to turn from a chance for economy into a risk for the whole community, we should "pace" ourselves.

We can notice the following phases of the development of a sustainable tourism (WTO, 1998):

- the first phase represents the decision to include in the tourist circuit a certain area and to erect the tourist constructions required for those tourist facilities;
- the second phase represents the progressive development of the tourist activity (along the responsibility for the environment protection and the compliance with the sustainable tourism);

During the first phase, if the tourist activities are carefully planned and performed, the environment problems can be solved since the same phase. For this, the selection of the area (of the ecotouristic management) is decisive to avoid the ulterior conflicts in the relation with the environment, such as:

- limitation of the damages brought to the landscape by: position of the resort, organization of transport, architecture, methods used to provide the facilities, etc.
- compliance with the responsibilities of the: local authorities, economic agents, local population (which must be consulted regarding the opportunity of the touristic project; it can oppose to it if it considers that its interest have been neglected.
- assessment of the impact over environment (according to the directives of the European Union), each member state having to introduce in its national legislation provisions regarding the impact of large tourist (or other) projects on the environment.

It is difficult to make a clear distinction between what is positive and negative in tourism development because often the short term impact is positive and the long-term is negative.

Let us consider a very attractive natural area (Iron Gates Natural Park). Tourists come attracted by the natural beauty, the richness of ecosystems, the uniqueness of endemism, but if this wealth is not "really" protected, the environment will degrade and lose its tourist destination attractiveness.

We must not forget that if the natural heritage of an area (even a protected area) attracts tourists, once it is degraded, tourists will leave as quickly, leaving behind a population that must bear the consequences due to the degradation of life environment, natural resources and tourism revenue decline.

An example of sustainable-environmental development of a valued tourist area in the Mehedinti county identity is (and hopefully will remain) **the Iron Gates Natural Park,** which reunites in its geographical area a series of "superlatives" among which:

- The Danube Gorge – the longest gorge in Europe (134 km);
- The largest Natural Park in Romania (115,655 ha);
- The largest hydroenergetic facility in Romania (the Hydropower and Navigation System Iron Gates I)
- The protected area with the greatest ethnic diversity in Romania
- Special geological and geomorphological diversity which may confer the status of outdoor geological museum;
- A high biological diversity – more than 1600 plant taxa (higher plants) and over 5200 fauna taxa;
- High diversity of plant associations, in this area being identified **171 associations**, of which **26** are **endemic to Romania** and **21** are **of community interest**;
- Presence of wetlands which are important habitats for protected bird species worldwide;
- Traces of human settlements from the Paleolithic, Mesolithic, Neolithic – evidence proving the area's living history: cities, monasteries, churches;
- Historical and architectural buildings, water mills unique as operating system;

A **National Park** is a protected area whose main purpose is the protection and preservation of landscapes created by the harmonious interaction of human activities and nature over time (Law no. 462/2001).

Iron Gates Natural Park objectives are: conservation of landscape features, biodiversity, ethnic and folk traditions and cultural values, development of harmonious relations between nature and society by promoting activities without impact on the environment and international cooperation in biological conserving of the Danube river basin.

Iron Gates Natural Park covers an area of **115,655 ha**. Located in the south-west of Romania, its space belongs to the counties of Caras-Severin and Mehedinti. The Natural Park' limits are the fairway of the Danube to the south, the river Nera in the west, the watershed directly tributary to the river Danube in the north (partially) and a sinuous line that starts downstream from Gura Văii to Motărăț peak the the east. This nature reserve, adjoining the Ecological Serbian National Park, on the Belgrade-Timoc segment – based on a joint program of cooperation with Serbia – would make in the region an ecological area on a large expanse, with major interest for the three neighboring countries and with economic and social implications very favorable for the Balkans, in the perspective of sustainable development (including tourism) in the EU.

Among the ecotourism activities to be held in the Iron Gates Natural Park the following can be mentioned:

- *mountain tourism* (marked tourist routes - Pemilor route, linking all the villages inhabited by the Czechs in the Iron Gates Park);
- *cruises on the Danube* (departing from Orsova);
- *scientific tourism* (for habitats and species of protected plants and animals);
- *speleology* (in the limestone areas of the Park: Cazanele Dunării, Coronini-Moldova Nouă-Gârnic, Sirinia)
- *birdwatching* (in the wetlands in the west of the Iron Gates Natural Park)
- *sport fishing* (on the Danube – catfish, carp, perch, starlet, etc., and on inner rivers – trout, barbel etc.).
- *traditional festivities and celebrations* (Neda, celebrations of minorities);
- *cuisine* (fish dishes, goat meat, dairy products, vegetables, sweets);
- *Danube water sports* (rowing, canoeing, jet ski);
- *race cycling and mountain biking;*
- *skiing* (on forest roads);
- *cultural sightseeing* (archaeological, religious);
- *visiting water mills* in the valleys Elişeva, Povalina, Camenita;
- *visiting museums and collections* (museums and ethnographic and religious collections)
- *visiting villages populated by Czech and Serbian ethnics.*

The progress of the touristic activities, the second phase of the development of a sustainable tourism represents the active implication of all interested parties (local suppliers of tourist services and local authorities, together with the local population) in actions meant to solve the environment problems using the economic or juridical means to force economic agents to use the environment protection equipment.

Finally, the contribution of tourism to sustainable development is part of a social and political process in progress. It also clearly shows us, on the one hand, the growing importance that tourism registers today, despite the global financial crisis (Mazilu, 2010a, 2010b, 2010c) and, on the other hand, the major stake represented and offered by this form of development – sustainable development – on all territories worldwide.

Given this complexity, but also to give a pragmatic sense to their intervention, communities, especially those at the regional level, have – perforce – entered in a specific process of territorialization of their policies. Each declines, globally or by specific sectoral policies, the priorities over a territory, whose functional area fluctuates from one community to another.

Harmonious development of tourism throughout the territory contributes to economic and social growth and to alleviate imbalances between different areas, constituting an important source of household income increase.

Tourism is a means of developing rural areas, by extending the area of the specific offer and creating jobs in rural areas, other than traditional ones, improving living conditions and increasing income of local people. With respecting and promoting the principles of sustainable development, tourism is a means of protection, conservation and capitalization of the cultural, historical, architectural and folklore potential of countries. By adopting a strategy of sustainable tourism development and enforcing environmental protection

Sustainable Tourism of Destination, Imperative Triangle Among: Competitiveness, Effective
Management and Proper Financing

109

measures, fundamental values of human existence (water, air, flora, fauna, ecosystems, etc.), tourism has at the same time an ecological vocation.

Global trends and priorities change: more than ever, the great challenge for the tourism sector is to remain competitive and sustainable by recognizing that, on the long term, competitiveness depends on sustainability.

It is in the interest of tourism to be active in the issue of sustainable development (Mazilu, 2007a, 2007b, 2007c, 2007d) and to cooperate with other industries in providing resource base quality and its survival.

We must not forget that it is essential that tourism be politically accepted as a priority, without compromising durability. Without support and political commitment to sustainable tourism, tourist programs based on the principles of sustainable development will not be implemented.

After a useful and necessary previous reflection on notional interpretations of the new type of tourism, and clarifying the multi-expressional use "for another type of tourism", which we use in various names such as: sustainable tourism, responsible tourism, social tourism, joint tourism, integrated tourism, fair tourism, community tourism, etc., the article analyzes the compatibility between tourism and the concept so addressed in literature, that of sustainable development and environment, the various constraints posed by financial return, the requirements of tourist market, gaps in the management of tourism resources and best practices to be established for tourism to become sustainable (Mazilu, 2006).

Since **sustainable tourism represents a goal,** one must understand that any type of development which includes development of tourism gives rise to certain changes in an area. However, these changes should be maintained within acceptable limits, so that sustainability to be achieved. Sustainable tourism can be best achieved through careful planning, development and proper management of the tourism sector based on some principles.

Consequently, there was the need for a new, professional leadership of tourism that attracts governments and private and public sector partners more, based on **principles** of sustainable development, namely:

- the environment has an intrinsic value which is especially great for tourism, which should be enjoyed by future generations;
- tourism must be seen as a positive activity that will benefit:
  - the environment
  - local communities
  - visitors;
- the relationship between environment and tourism can be developed so that the environment sustains long-term tourist activity, tourism, in turn, being "obliged" not to cause environmental degradation through its unfolding.
- development of tourist activity must comply with environmental, social, economic, cultural features of the geographical space it takes place in;
- the goal of tourism development must always be the balance of the needs of tourists with their destinations and hosts;

- tourist industry, governments, authorities responsible for environmental protection and international bodies must respect these principles and work to implement them.

Avoiding the trap of "preconceived ideas" (which often seem to be false) like sustainable development is strictly the "business" of industries that process. Why fake? Because global warming and pollution directly relate to the tourist phenomenon: climate change, for example, have an impact in season changes, so have a direct incidence in the seasonality of the tourist phenomenon, the freeze-thaw phenomena damage, they damage infrastructure, the recent, highly publicized floods in exotic areas of high tourist attraction simply alter tourist experiences in this environment.

- **Sustainable tourism is not and should not be just a "panacea" of governors**. Sustainable tourism concerns everyone from:
  - **various governmental levels**: regulations, landscape protection, legislation, etc.
  - **associations and NGOs in the field of tourism**: sustaining the tourism phenomenon, examples of good practice, environmental protection animation, etc.
  - **industrial objectives** - to adopt new non-polluting practices and environmental protection;
  - **tourists** to know and apply the ethical code of tourists, to be responsible towards the environment and tourist destinations, to highlight their value;
  - **local population**: the need to show hospitality, knowledge of the tourist code, it being involved itself in activities that safeguard and enhance the environment;
  - **up to all components of tourist industry**: from tourist destination, regardless of size, to the types of tourism.
- **Sustainable tourism** seeks not only environmental protection. When we say sustainable development we mean impacts (either environmental or socio-cultural, etc.), and sustainable development requires sustainable management of these impacts. It is a matter of achieving balance and harmony in a spirit of sustainability (even thrive) on the long term.
- **Sustainable tourism is a tourist product** that can be sold to tourists. Sustainable tourism is a way to design, plan and manage sustainably tourism activities. At the same time it requires a change in the style of management, behavior, attitudes and habits.

There are no success "recipes" in the development of a sustainable tourism, but we can take into account several "tracks" of intervention (Baker, M. J. (2008):

- promoting an action plan in partnership with the "key" players in the development of the tourist phenomenon, including the involvement of industry leaders;
- demystifying the meaning of sustainable development;
- communication in sustainable development;
- integration of sustainable development as a factor in the capitalization of the tourist industry and betting on the strong links between quality and sustainable tourism;
- integration of sustainable development into the training of future "managers" of the tourist phenomenon;
- recognition and promotion of successful cases of sustainable development;
- tourism education in the spirit of sustainable development to make the best choices in the development of the tourist act.

4.3 Fields of action of players in sustainable tourism

Based on the principles of the sustainability triangle and on the related fields of action, against which interested players respond, the discussion on sustainable tourism is emerging. The themes belonging to the main domain of sustainable tourism or "sustainable tourism development" can be actually divided into subdomains.

We present a non-exhaustive list of fields of action in figure 8.

All players must complete a specific task to give the force required for sustainable tourism development and for each of them to obtain benefits by keeping an attractive tourist environment that protects at the same time the environment (Pârgaru, 2009).

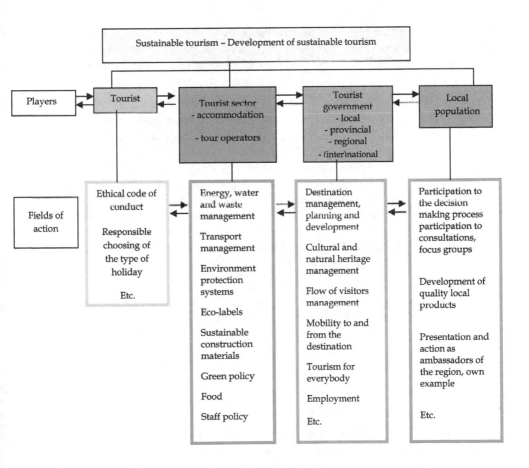

Fig. 9. Fields of action of players in sustainable tourism

Sustainable tourism development requires more and more knowing participation and involvement of all players involved, and a strong political support to ensure broader participation and creating a general consensus in the field of tourism development policy.

Sustainable tourism is the fruit of continuous efforts and requires constant control of desired or undesired effects of this activity, which includes the adoption, whenever appropriate, of preventive measures or that correct certain imbalances.

Tourism experts have decided, therefore, to lean on quality, sustainability and modernization of tourist services. Concretely, they will study the links between competitiveness, skills and information technology in this field, developing suggestions to improve regulation in this sector, focusing on the famous "services" directive. Finally, they will even try to put into place a sustainable development strategy in this so exposed area to polemics.

Local people, tourists, authorities responsible with tourism management and the private sector are the key players. The manner in which they can provide the necessary dynamics to this process depends on their own goals and vision of the outcome, the structure of ownership, their financial means and the influence they exert. We refer to the following activities to ensure sustainable tourism:

- in relation to the implementation of sustainable tourism, tourists pay increased attention to contact with the local population and show a more respectful attitude towards nature, environment and the surrounding areas;
- the tourist sector invests especially in its economic capital, accomplishing in this way especially technical improvements of the natural environment; however, this does not mean little attention paid to human capital (e.g., staff);
- authorities managing tourism have a central, very important role in the management of tourist destination and thus the development of tourist related products, while having the responsibility to promote proportionally the environment, population and economic significance that tourism offers to the region, they must share these resources in all fields;

On the other hand, sustainable development can be revealed to us as an alibi, even a utopia maintained voluntarily by political and economic players, anxious to legitimize and to prove their economic logic but also to preserve the economic strengths already acquired. We believe that to resolve this situation it would be enough to adapt socio-economic and ecological systems of the "given" of globalization: **Sustainable Development.**

4.4 The support capacity of a destination

Support capacity is used in the context of sustainable tourism. It is often proposed as a method for assessing the intensity of tourism development that can be supported by a tourist destination taking into account the economic, environmental and social characteristics of that destination.

- In literature, an interesting definition of this concept was formulated:

Sustainable Tourism of Destination, Imperative Triangle Among: Competitiveness, Effective
Management and Proper Financing

113

- „The maximum number of visitors that can be found in the same time in a tourist destination, without causing adverse effects on the physical, economic and sociocultural aspects of the society / community and without causing a (significant) decrease in the level of visitor satisfaction (Clivaz et al., 2004).

- The last part of the definition, in particular, provides a unique tool for measuring and monitoring the support capacity as visitor satisfaction, among others, is closely linked to the quality of service provided and quality can be measured in a simple way.
- In destination management, the simple question "how much is too much?" cannot be answered easily. A destination is a dynamic identity that the public sector must constantly adapt according to the possibilities it has and the impulses it reacts to. Choices must be made together with local players, but during this process of choice of responses and reactions to external elements the previous limit of the support capacity may be exceed. For example, a destination can receive more visitors and, at the same time, provide better living conditions than an appropriate traffic management plan has designed. This traffic plan can be useful both for tourists and locals.
- Due to the complexity of the development and management policy, the support capacity must evolve from a unique concept (limit of the maximum number of visitors an area can receive at a given time) to information/monitoring system that supports decision making.

Using methods that belong to the spatial sciences and geography, tourism economy, the reception capacity of a territory can be determined, respectively its ecological support capacity in relation to natural and human resources and the space on which it materializes. This renders the number of tourists that can be received, at a given time, on a territory or in a resort, without this tourist flow and related facilities (reception equipment or tourist production) to harm the environment through its degradation or of tourist resources.

Following the development of environmental science, of the increase of the amount of information on economic and social sciences, of the deepening of research in these fields, their connections with the tourist industry have led to the delineation of the following types-support capacity for tourism:

Ecological capacity refers to the establishment of that level of development of structures and tourist activities without strongly affecting the environment through the degradation of its components. This also applies to natural components (air, water, soil, vegetation, and fauna) and the production process and economic recovery, that do not involve special investment costs caused by the degradation of a tourist destination;

Physical capacity has an essential role in determining the saturation level that tourist activities can reach, beyond which environmental issues begin to emerge. The ascending development of tourism, manifested in recent decades raises the question of the emergence of many forms of pollution (coastal, of mountain areas, etc.). Protecting the physical components of the territory is made through investment in technology and by providing high quality level of tourist services;

Social-responsive capacity aims the importance of preserving good relations between hosts (indigenous population) and visitors (tourists). Since the moment that local population finds that tourist activities also contribute to environmental and cultural degradation, hostile, rejection reactions may arise, also a reduction of the threshold of tolerance being recorded. To avoid such situations, the development of tourist areas or places should take into account the traditional life of the inhabitants, their customs, etc.

Economic capacity highlights the capitalization of all resources present by tourist activities and it represents the capacity to maintain the tourist function of a given area. Operating efficiency is measured by the ratio of costs and benefits, and the share of benefits can be increased by using new technologies. The level of costs is also given by the "qualitative and quantitative value of resources (natural, cultural, labor, general infrastructure, etc.);

Psychological capacity is related to the negative perception of tourists towards the tourist destination, following environmental degradation or poor attitude of the local population.

This concept is attached to supporting tourist motivations for a particular destination and maintaining their own personal satisfaction. Its application is also conditioned by the quality of management activity that can lead, ultimately, to the loyalty of demand.

All these different types of support capacities, in close relation with tourist activities, determine the tangible or intangible limit, measurable or immeasurable, of an area which has or which may be assigned a tourist function.

Although these capacity indicators do not provide a standard formula, due to the fact that some components of the natural or cultural area are difficult to quantify by series of statistical and mathematical data, however, these support concepts give us the measure of the sustainable development of tourism. At the same time, the types of support capacities also indicate the extent to which the impact of level of tourism may have on the environment, allowing the possibility to identify ways to reduce the degradation caused by the movement and tourist activities.

5. Conclusion

Imperative mutations that must be reflected in the policies of post-economic crisis revival in the tourism industry are:

- **The focus should move on general economic development**, rather than uncoordinated and inconsistent measures targeted for tourism industry
- **Promotion of a policy in favor of the tourist**, instead of the priority that is granted today to tour operators and - to a lesser extent - the local tourist accommodation industry
- **Improved communication through more intelligent branding actions and advertising**
- **A networking industry par excellence, Romanian tourism can achieve even more than other areas of a country's economy.**
- **Tourism has been defined as a system in which interdependence is essential and collaboration and cooperation between different organizations within a tourist destination creates the tourist product** (Mazilu, 2009).
- **Under the impact of globalization,** the development of tourism in each country of the world becomes possible only under the conditions of an optimal public-private partnership;

Sustainable Tourism of Destination, Imperative Triangle Among: Competitiveness, Effective
Management and Proper Financing

115

- **There is no real public-private partnership** in the field, and the institutions empowered to create a certain education and behavior to support the sustainable development of the Romanian tourism do not make the necessary efforts;

Basically, in this crisis, Romanian tourism should maintain its accommodation capacity and improve services (Mazilu M.E, 2011). The real benefits will be obtained later. The financial and economic crisis effects are felt globally by all players in the market, regardless of the operating field.

6. References

Ambiehl, C. ; Aujaleu, C. ; Galienne, G. (2002). *Le marché du bien-être et de la remise en forme avec l'eau Agence française de l'ingénierie touristique*, Consulting, Agence, p.18

Ashworth, G. (2001). The Communication of the Brand Images of Cities, paper presented at the Universidad Internacional Menedez Pelay Conference: *The Construction and Communication of the Brand Images of Cities*, Valencia Spain and *From City Marketing to City Branding: Towards a theoretical framework for developing city brands*, Place Branding, Vol. 1, No.1, pp. 58-70.

Baker, M. J. (2008). Critical success factors in destination marketing, in *Tourism and Hospitality Research*, Vol. 8, No. 2, Palgrave Macmillan, pp.79-97.

Bessy, O. (2008). *Sport, Loisir, Tourisme et développement durable des territoires*, PUS, Cedex, p. 44

Cândea, M.; Stăncioiu, F.A; Mazilu M.E; Marinescu R.C. (2009). The Competitiveness of the Tourist destination on the the future Tourism Market, in *WSEAS Transactions On Business And Economics*, VOL.6, Issue 7, August,2009, ISSN: 1109-9526 pag.374-384

Clivaz; Hausse; Michelet (2004). *Tourism monitoring system based on the concept of carrying capacity–The case of the regional natural park Pfyn-Finges (Switzerland)*, document of the Research Institute Finnish Forest, 2004

Connell, J.; Page, S.; Bentley, T.(2009). *Towards Sustainable Tourism Planning in New Zealand: Monitoring Local Government Planning under the Resource Management Act*, ElsevierJTMA-D-08-00172R1

Constantin, D.L. (2000). *Introducere în teoria şi practica dezvoltării regionale*, Editura Economică, Bucuresti, România

Echtner, C.; Brent Ritchie J.R. (1991). *The meaning and measurement of destination image*, Journal of Tourism Studies

Gartner, W.C (2009). Image Formation Process, in *Journal of Travel and Tourism Marketing*, 2(2/3), pp.191-215.

Go, F.M.; Van Fenema P.C. (2006). Moving Bodies and Connecting Minds in Space: It is a Matter of Mind over Matter Advances in Organization Studies 17, pp. 64-78,

Govers, R.; Go, F. (2009). *Place Branding. Local, Virtual and Physical Identities, Constructed, Imagined and Experienced*, Palgrave Macmillan, London.

Gowers, R.; Go, F. (2009). *Place branding, local, virtual and physical identities, constructed, imagined and experienced*, Palgrave Publishing House.

Ionescu, I. (2000). Turism, fenomen social-economic si cultural, Editura Oscar Print, Bucureşti.

Lindstrom, M. (2009). *Branduri senzoriale. Construiţi branduri puternice folosind toate cele 5 simţuri*, Editura Publica, Bucureşti

Low no. 462/2001 on Protected Natural Areas, natural habitats, flora and fauna, http://www.cdep.ro/pls/legis/legis_pck.htp_act?ida=30639

Luca, C. (2007). The relationship between social dominance orientation and gender: The mediating role of social values. *Sex Roles*, 57, 159-171.

MacKay, K.J.; Fesenmaier, D.R. (2000). An Exploration of Cross – Cultural Destination Image Assessment, in *Journal of Travel Research*, 38(4) May, pp. 417-23.

Mazilu, M. E. (2006). *The Touristic management from an ecological perspective*, publicat în Analele Universităţii din Craiova, Seria Ştiinţe Economice, Anul XXXIV, Nr.34, Vol.1

Mazilu, M. (2007). *Le tourisme roumain dans le contexte du tourisme europeen*, Universitaria Publishing House, Craiova, p.82.

Mazilu, M. (2007). *Tourist Geography*, Didactical and Pedagogical Publishing House, Bucharest, , p. 234.

Mazilu, M. E. (2007). The ecological component of the lasting development, Conferinţa Internaţională B.EN.A: *Sustainable Development in Balkan Area: Vision and Reality*, Alba Iulia, 18-20 iulie 2007, publicat în *Journal of Environmental Protection and Ecology JEPE*, Book 1, vol. 10, ISSN:1311-5065, pag.131-136,2009.

Mazilu, M. E. (2007). Turismul - o relaţie privilegiată cu dezvoltarea durabilă, articol publicat în *Revista de Marketing On-line*, Nr.4, volumul1, Editura Uranus, Bucureşti , ISSN:1843-0678, pag.64-70

Mazilu, M. (2008). Un Tourisme fait pour durer, in Vol. and the program Colloque international: *Services, innovation et développement durable*, Poitiers, France file://F:/Colloque SIDD – Poitiers/ Communication/Mazilu Mirela.pdf. ,p.33.

Mazilu, M.E. (2008). *Competitivité et excellence dans l'aménagement touristique durable du territoire*, Les Annales de l'Universite Valahia de Targoviste section: Sciences Economique, anul XIV, nr. 26

Mazilu, M., Marinescu, R. (2008). Sustainable Tourism in Protected Areas – Case Study of the Iron Gates Natural Park, *Rural Futures Conference*, organized by University of Plymouth and School of Geography, Plymouth, the Great Britain, ISBN: 978-1-84102-185-0, p. 1-7.

Mazilu, M.E. (2009). *Actorii implicaţi în turismul durabil al unei destinaţii*, publicat în Proceedings of The 2-nd INTERNATIONAL TOURISM CONFERENCE: *"Sustainable mountain tourism - local responses for global changes"*, Drobeta Turnu Severin, Editura Universitaria, Craiova, ISBN: 978-606-510-622-2, pag.215-221

Mazilu, M. E. (2010). Key elements of a Model for Sustainable Tourism, in *International Journal of Energy and Environment*, Issue 2, Volume 4, ISSN:1109-9577, p45-54.

Mazilu, M. E. (2010). Opportunities and Threats for Romania as a Tourist Destination after the World Economic Crisis, in *Proceedings ISI of 5-th WSEAS International Conference on Economy and Management Transformation(EMT'10)*, ISSN: 1792-5983, ISBN:978-960-474-240-0, West Timisoara University, pag.66-72.

Mazilu, M. E. (2010). Towards a model of an optimal-sustainable tourist destination, in *ISI Proceedings of International Conference: Cultural, Urban and Tourism Heritage, CUHT2010,Corfu,Greece,24-26,Iulie,2010,   ISBN:978-960-474-205-9,   ISSN:1792-4308, pag.28-35.*

Mazilu, M.E.; Marinescu R.; Sperdea N. (2010). The Quality of Tourism Services under the sign of sustainability, *publicat în Analele Universității din Oradea, Științe Economice, Tom XIX, Issue 2, ISSN:122569, p.992-999*

Mazilu, M. E. (2011). Romania – an Attractive Tourist Market after the World economic Crisis, in *International Journal of Energy and Environment,*Issue 2,Volume 5, p.212-221,

Morgan, N.; Pritchard, A. (2004). Meeting, A.; Pride, R. (eds), (2004). *Destination Branding: Creating the unique destination proposition,* 2nd Edition, Oxford: Butterworth-Heinemann.

Nistoreanu (2005). The trilateral relationship ecotourism - sustainable tourism - slow travel among nature in the line with authentic tourism lovers, *Journal of Tourism,* No.11, p35

Pârgaru I.; Mazilu, M.; Stăncioiu, A.F. (2009).   Tourism – „Energy Source" for the Sustainable Development, publicat în *Metalurgia International,* vol. XIV, no, 8, Special Issue , Editura FMR ISSN 1582-2214, pag.64-68

Rondelli, V.; Cojocaru, S. (2005). *Managementul calității serviciilor din turism și industria ospitalității,* Editura THR-CG, București, p.165.

Sanagustín Fons, M.V.; Moseñe Fierro, J.A.; Gómez y Patiño M. (2011). *Rural tourism: A sustainable alternative,* p. 552

Sorokin, P. A. (1967). *The sociology of revolution. New York,* NY: Howard Fêting, Inc.

Stăncioiu, F.A. (2003). *Marketing turistic,* Editura SITECH, p 104

Stăncioiu, A.F.; Pârgaru, I.; Teodorescu, N.; Talpaș, J.; Răducan, D. (2008). Imaginea și identitatea - instrumente de poziționare în marketingul destinației, lucrare susținută la Sesiunea de comunicări științifice „*Cercetare interdisciplinară în turismul românesc în contextul integrării europene",* organizată de Institutul Național de Cercetare și Dezvoltare în Turism, Ighiu

Stăncioiu, A.F.; Pârgaru,I.; Mazilu, M.E. (2009). Brandul destinatiei, căteva repere conceptual metodologice în marketingul destinației, in *Proceedings of The 2-nd International Tourism Conference:"Sustainable mountain tourism-local responses for global changes",* Drobeta Turnu Severin, Editura Universitaria, Craiova, ISBN: 978-606-510-622-2, pag.283-289

Stănciulescu, G. (2004). *The Sustainable Tourism Management in the Urban Centres,* The Economic Publishing House, Bucharest, p. 23.

Tao T.C.H.; Wall, G. (2009). Tourism as a sustainable livelihood strategy. *Tourism Management, Volume 30, Issue 1, February 2009, Pages 90-98*

World Tourism Organisation, WTTC, The Earth Council. (1995). Agenda 21 – for the Travel and Tourism Industry: Towards Environmentally Sustainable Development

World Tourism Organisation (1998). Guide for Local Autohrities on Developing Sustainable Tourism

World Tourism Organisation / UNEP. (2007). Making tourism more sustainable, A guide for policy makers, in the *Sustainable Tourism Group report*, "Action for more sustainable European tourism"

www.ecotourism.org

6

Built Heritage and Sustainable Tourism: Conceptual, Economic and Social Variables

Beatriz Amarilla and Alfredo Conti

Research Laboratory on the Territory and the Environment (LINTA)
Scientific Research Commission, province of Buenos Aires (CIC)
Argentina

1. Introduction

The origin of tourism in the modern world is closely linked to recognition and appreciation of cultural heritage. In the eighteenth century, the so-called "Grand Tour", from which the word *tourism* comes, consisted of trips of intellectuals and artists from different countries of Europe to the Italian peninsula, and especially to Rome, to take direct contact with relics of classical antiquity. The ancient world, especially the Roman civilisation, had been "rediscovered" by the culture of the Renaissance and Rome had become a destination for those who wanted to know and appreciate classic art, which was taken as a source of inspiration for artistic production of the time (Choay, 1992). This habit meant the implementation of infrastructure and equipment to meet the needs and requirements of the travellers, including transportation systems and accommodation, all of which constitute a background of modern tourist facilities. This initial form of tourism included some components of its current definition: the idea of "tour" meant that travellers returned to their places of residence once their expectations had been met, they were motivated eminently by cultural purposes and funds invested in the destination came from their home countries. In short, it was, in modern terms, a practice of cultural tourism, reserved for a selected social group in terms of education and economic position.

The major economic, social and cultural changes induced by industrialisation all over the world took to new modalities of tourism; among the rights recognised to workers, free time, vacations and leisure appeared, especially throughout the twentieth century New social groups were gradually incorporated to the practice of tourism. At the same time, new heritage items expanded the realm of tourist attractions, something that has occurred up to the present time. The interest in nature typical of the scientific field between the seventeenth and eighteenth centuries was transferred to the realm of art especially by the nineteenth century Romanticism; the natural environment was considered a source of contemplation and relaxation for body and spirit. The translation of this expansion of the heritage concept impacted in the field of tourism, besides sun and beach and cultural tourism. Nowadays, several modalities appeared, among them ecotourism, adventure, health, business, religious, gastronomic tourism, etc. Over the twentieth century, and particularly the period after the World War II, the progress in transportation and general improvements in revenues facilitated the access of new social groups to the possibility of travelling. Social tourism emerged, which assured the possibility of holidays for workers, and mass tourism, which has continued to expand, making tourism a top economic activity at international level and, for many countries and regions, a source of economic growth and overall development of communities.

These concepts have undergone several changes throughout time. The qualitative and quantitative evolution of tourism as economic and social activity coincides with the development of heritage concept, assessed mainly in the late twentieth century. New categories of goods would be added to the initial idea of "historic monuments", which according to the Venice Charter (1964), these are testimonies of *"a particular civilisation, a significant development or a historic event"*; the concept of monument does not longer refer only to great creations but also to *"more modest works of the past which have acquired cultural significance with the passing of time"*. In the late twentieth century, new heritage categories were considered; in 1992, UNESCO introduced the concept of cultural landscape, consisting of the joint work between man and nature. In 2005 the notion of heritage routes was included in the *Operational Guidelines for the implementation of the World Heritage Convention*, which implies the consideration of *"tangible elements of which the cultural significance comes from exchanges and a multi-dimensional dialogue across countries or regions, and that illustrate the interaction of movement, along the route, in space and time"*. Intangible heritage, consisting of literary and musical works, traditions, social practices, oral history, gastronomy and traditional knowledge gained significant ground in the theoretical debate and in the field of heritage management, which is manifested in the adoption by UNESCO of the Convention for the Safeguarding of Intangible Heritage in 2003. In this framework, historic towns and urban areas became one of the most significant heritage assets, since they express, perhaps better than other heritage categories, all the complexity of human relationship with the environment, merging tangible and intangible heritage components. Nowadays, heritage is recognised as social construction and it is highlighted the active participation of all social actors in its identification and management, when defining heritage, according to Prats (1997), as *"the symbolic referent of the cultural identity of the community."*

These two concepts in evolution, heritage and tourism, are linked with the formal appearance of the concept of sustainable development in 1987. Its more well-known definition indicates that it deals with a development that has to satisfy the current needs without threatening the ability of future generations to solve their own needs. Though the term "sustainable" may be mainly associated to natural resources, a public policy of sustainability cannot exclude today conservation, management and use of the built heritage.

Tourism has had a global booming growth during the last decades, and the close relationship between heritage and tourism brings about opportunities and threats. Among the advantages we can mention the attainment of economic resources, creation of jobs, provision or improvement of facilities and the urban infrastructure, enhancement of public spaces and building restoration as well as the consolidation of the local identity. However, there are important threats related to sustainability. These not only imply alteration or destruction of material components of buildings and sites but also they may distort their values and meanings.

As regards this issue, key economic factors, which need to be identified and analysed, remain. Tourism has always been studied as an economic activity, measuring its incidence, for instance, in the national gross product. However, when we include heritage, the economic study is subtler and less spread. The International Council on Monuments and Sites (ICOMOS) adopted an International Charter on Cultural Tourism in 1976. The extraordinary growth of tourism over the last decades of the twentieth century led ICOMOS to review the Charter; a new text was adopted in 1999, more suitable to the demands of the moment. This text introduces concepts and recommendations related to the proper interpretation and transmission of heritage values and meanings, with the need to consider the tourism use of heritage as a tool for the integral development of host communities through the idea of participatory planning involving all stakeholders.

In the context above mentioned, the main purpose of this chapter is to study the relationships among three key concepts: built heritage, cultural tourism and sustainability. The conceptual, economic and social variables of this issue are particularly emphasised, making special mention to the case of Latin America countries.

The main problem that will be discussed in this chapter is the degree of conflict among these three concepts, which arise particularly when we pass from theory to practice. We will follow the next steps in order to analyse this problem and to think about the compatibility among sustainability, tourism and heritage. First of all, it is necessary to study the definitions and nuances of the sustainability concept, relating them to the updated conceptions of tourism and built heritage, according to current specialised bibliography and technical documents. Secondly, it is required to link the knowledge cited above to social and economic variables, factors that, according to our hypothesis, can cause the main distortion when we pass from theory to practice. Finally, it is essential to verify these through some case studies. In relation to that, we think that Latin America World Heritage towns are very appropriate examples in order to test the ideas developed before.

The analysis of successful experiences and of problems detected in many cases of historic centre rehabilitation of Latin America included in the list of UNESCO World Heritage – Cartagena de Indias (Colombia) and Colonia del Sacramento (Uruguay), among others – serve as guide for the development of a series of recommendations intended to achieve the tourist sustainable use of the cultural heritage, with social, economic and cultural advantages for the communities involved.

2. Sustainability and built heritage

2.1 About the sustainability concept

"Over the last thirty years, the concept of monument has grown, evolving from the individual building to the historic district of the cultural landscape. Slowly but surely, we will reach the concept of ecosystem, in which it will become obvious that the preservation of a site, even as a historic city, can occur only if it is possible to preserve its environment and all the activities that have traditionally supported the life in the site. Why preserve the fisherman's town if the river is allowed to go dry or if the industrial plant upstream is allowed to pollute the river and kill the fish?" (Bonnette, 2001).

The previous observation has the advantage of building a bridge between concepts which are very widespread, such as sustainable development and heritage conservation, but they often appear empty of content, or they are interpreted in a different way from the view of other disciplines, or they are not integrated according to a systemic view. Sustainability is a concept which has been applied, not very accurately, among the people responsible for policies and decision making (Meppem, 1998). There is no coincidence about its meaning, either.

If the sustainability definitions are considered, it may be concluded that most of them are expressed in normative or positivist terms. According to Keynes, a regulatory science can be defined as a body of systematised knowledge referring to the criteria of what something must be; instead, the positivist science considers what things are (Keynes, 1890, as cited by Meppem, T. & R. Gill, 1998).

The most widespread regulatory definition of sustainable development is that emerging from the so-called Brundtland Commission in 1987, which states that development must answer to the current needs without compromising the ability of future generations to meet their own needs. The success of this definition is sometimes attributed to its ambiguity.

In general, there is some agreement in the fact that the scientific positivism is incapable of answering by itself the political and cultural variables that lead the action to a sustainable development. The evidence resulting from the science domains, from sociology, philosophy, economy and law suggests that the conventional regulatory-positivist approach is not suitable from the epistemological point of view. The alternative is to develop a process in which it is taken into account the socio-cultural context where the environmental and economic information circulates, considering the development scenarios more completely.

According to Pearce, the capital stock takes different shapes: works and products, human capital (knowledge, abilities), natural, environmental or social capital. The latter involves the group of social relationships that produce welfare directly or indirectly. The built heritage, which is the main subject of our study, can be thought also as a capital, which comprises a stock of physical works along with ideas, beliefs and values that gather communities and link the present with the past (Pearce, 1998).

As a general rule, the sustainable development requires that the capital stock transferred to the next generation is lower than the capital stock in hands of the current generation. From that, two new concepts have arisen:

-   Weak sustainability: the stock should grow through time, but its composition is irrelevant. For instance, if the environment is now degraded, this can be justified since the benefits resulting from the proposed activity are higher than the cost of the resulting degradation. Many projects of conventional urban construction and intervention can be included within the weak sustainability concept.
-   Strong sustainability: some goods, like those included in the natural or cultural heritage, are so important that it is indispensable its conservation. Some of the reasons for this are the following:
    a.   The environment has intrinsic values and it is not replaceable. There is no possible replacement (irreversibility).
    b.   The value of the environment and its components is uncertain, therefore, it should not be destroyed for caution. This reason may respond to a kind of "non-use values" (for instance, a plant species can be the key in the future healing of a disease, and this would be impossible if this species becomes extinct).

From the social point of view, the strong sustainability is related with the idea of investing in activities which enable social progress, that is to say, improving the public participation, democracy, the reinforcement of local communities, information flow and the human capital, since the formation of human capital through education enables the development of other social values. Within the strong social sustainability, it is possible to focus on the cultural capital and, specifically, on the built heritage. There is some agreement about the fact that art and architecture have values similar to those "intrinsic values" of the natural environmental ones. In fact, according to Pearce, it is about a modern interpretation of Ruskin's thought:

"... it is not about convenience or feelings when trying to preserve or not buildings from past times. We do not have rights to touch them. They are not ours. They belong partly to those who built them and partly to all mankind generations that will follow us." (Ruskin, as cited by Pearce, 1998).

From the point of view of weak sustainability, this position can be objected. It is held that conservation at any cost is not practical and even doubtful from the moral point of view since the resources spent on conservation could have been used today with other purposes, maybe with higher benefits. The defenders of sustainability often disregard the basic principle of economy: the opportunity costs. Beyond the moral arguments of conservation, the financial resources spent on conservation, especially in developing countries, can be used to solve present basic needs (food, health, housing), which can also be considered as "rights".

It is also certain that not all the present goods can be preserved with the expectation that future generations consider them as their "cultural heritage". Leaving the decision in hands of "specialists" would be in agreement, perhaps, with Ruskin's thought, but this attitude can be considered inappropriate according to the present concept of heritage as "social construction".

2.2 Man, nature and culture

Each generation has a capital that mainly includes three kinds of goods: natural, cultural (personal property and real estate) and human resources. There is a share, quantitatively smaller, of natural and cultural goods, considered as having special features, and because of this, they deserve to be protected, so that they can be enjoyed by the present and next generations (Lichfield et al., 1993). This apparently simple issue, in practice, results in countless difficulties of complex order, among which economic aspects play a key role.

In the previous paragraph, it is possible to foresee two topics that deserve consideration: that of the relationships among man, nature and culture; and heritage conservations as a problem that affects or involves different generations (the latter, key point in the classical definition of sustainability).

Regarding the first issue, it can be stated that until mid twentieth century the human myth of the supernatural man prevailed, and the opposition between nature and culture was the basis for the prevailing anthropological model. The world seemed to be built by three overlapped strata, isolated from each other: man-culture, life-nature, and physics-chemistry. This situation began to be modified in the 1950s with the opening of the gaps between these tight paradigms, enabling a new concept in the relationships between the natural and cultural things.

It is owed to Schrödinger, pioneer of the biological revolution, the main idea that: living beings are nourished not only by energy but also by complex organisation and information. Thus, the human society, which can be considered the most emancipated as regards nature, actually nourishes its independence from multiple relations: the more autonomous the living system, the more dependent on the ecosystem where it is integrated (Morin, 1973).

This complexity sometimes seems not to be taken into account in the conservation environment of the natural and cultural heritage, and this can result in the fact that the implemented measures do not produce the expected results. For instance, many of the national and international efforts have been centred on the creation of national parks and protected areas, aimed at the conservation of pristine and intangible goods. Considering cultural landscapes is relatively recent, understanding as such those where the physical and biological features have been extensively modified by human activity. This means that the decisions and social - economic processes prevail in determining spatial patterns and landscape characteristics. Disregarding this aspect, sometimes the protected areas do not offer an integral solution for the wild life conservation, aim of its creation. Then, it becomes necessary the programs oriented specifically to the needs of rural residents, who live on agriculture and on the exploitation of the wild life in the area. This kind of approaches to the conservation fulfils a double purpose: to protect the wild habitat and to respond to human needs at the same time (Young, 1997:137).

Regarding the problem between generations, it is known that, in the Earth Summit held in Rio de Janeiro in 1992, the sustainable development arose as one of the most urgent and critical topics of international politics. Some specialists consider that this critical issue has been mostly dealt from an "emotional" point of view, but not enough progress has been made in the construction of sustainable patterns for a modern industrial society. A serious position requires the study of the resource assessment, what usually results from its contribution to use or profit. The key factor is how the value is considered so that present and future have an equal treatment. Sensitivity is necessary in relation to both and this implies a symmetrical treatment of the generations, in the sense that neither the present nor the future must profit at the expense of the other. Thus, there is no preference for the romantic vision that privileges the future or for the consumption view that only decides in function of the present (Chichilnisky, 1997).

2.3 Tourism and sustainability

Tourist resorts attract cyclic populations throughout time. The tourist's perception of the social and environmental quality of the site influences the future attraction that such places bring about. If tourists have a positive perception of the site, an increasing number of people will visit the area. As this figure increases, the effects of tourism on social and environmental quality increase. Thus, degradation levels may be reached because of excessive visiting, and these levels can revert the tourism trend in the area (Lawrence, 1994:265-66). This process is known as "tourist cycle", shown in Figure 1.

Tourism is cyclic by nature, so are its social and environmental impacts. In Figure 1 there are two representative bands of the tourist population and of sustainability. The first grows through time up to an MTP (maximum tourist population) point, from which the people flow tends to decrease, as a consequence of excessive visiting, pollution, crimes, etc. On the other hand, the negative impacts that indicate a decline of sustainability take some time to revert, since a period of time is needed for the environmental improvement and socio-cultural changes. If the tourists' arrival continues decreasing, the negative impacts increase from a CNIT (change of the negative impact trend) point. Like in the supply and demand curves of economic models, the tourist population and sustainability flow around the intersection area of both bands. Such area is the sustainable development zone. If the tourist population and the negative impacts do not exceed the maximum levels that this area determines, a sustainable development can be reached in the long term.

Many tourist projects have ignored these tendencies, encouraging tourism growth beyond the recommended limits. This, in the long term, has resulted in the opposite expected effect: the decrease of the tourist flow together with social and environmental negative impacts in the area, sometimes very difficult to revert.

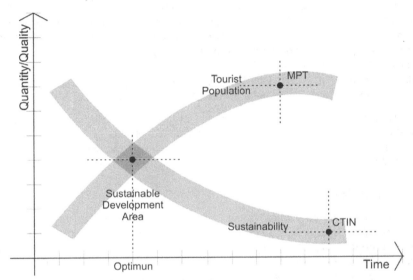

Fig. 1. Sustainable development area. Adapted from Lawrence (1994)

2.3.1 About ecotourism

The interest for the environment, developed in the industrialised countries since the 1980s, has encouraged particular ways of tourism, such as ecotourism. This fact contains potentially excellent opportunities of using tourism as protection tool of natural ecosystems, giving them a socio-economic value in their original estate.

The term "ecotourism" seems to have appeared in Mexico, chosen by Ceballos-Lascurain to define activities related to trips to natural areas in relatively pristine conditions, in order to study, admire and enjoy the flora, fauna and cultural expressions that may have existed (Filion et al., 1994). This concept has been broadened throughout time, including the environmental and socio-cultural consequences resulting from the tourist activity. Thus, nowadays, it is considered that ecotourism is a responsible trip, aimed at preserving the natural environment and at sustaining the local population welfare.

The definition of ecotourism includes a wide range of activities. Some market segments are small and well-defined (ornithologists, observation of rare species), while others are just the opposite (enjoying nature in trips mainly organised with another purpose).

Ecotourism seems especially positive in developing countries, since many of them are characterised by their biodiversity richness, but sometimes they do not have the necessary means for preserving the environment, favouring other forms of unsustainable economic development. In countries with a particular natural and cultural heritage, and with a special sensitivity or tradition in its conservation, these activities may become the main income source of the nation. In Nepal, "*the economic dream of conservationists*" seems to come true (Wells, 1994): a net of well-established protected areas produces, thanks to foreign visitors, an important part of the income of one of the poorest countries in the world.

In some developed countries, ecotourism has a significant importance. For instance, in Canada, millions of inhabitants perform tourist activities related to wild life, such as bird watching. These activities produce a high economic impact on the gross net product, on the income level of the population, on job creation and government income through taxes. In this country, an important part of the income the government receives through taxes from local ecotourism is invested in preserving wild life. Therefore, quantifying the socio-economic importance of ecotourism is a key factor for encouraging governments and companies to increase their efforts in environment conservation.

Though travelling with the purpose of enjoying nature has important benefits, there are also great conflicts. Beyond the well-known negative impacts on society and environment, there are other problems in the developing countries, such as money leakage: international airlines, hotel chains, etc. (Filion, 1994).

2.3.2 Cultural tourism and built heritage

Built heritage must also be studied as the basis for the development of sustainable tourist activities. According to "The Royal Heritage Site Working Group", tourism and heritage are activities that must find a balance, including understanding the permanent value of the sites with historic and heritage relevance, as well as respecting the original nature and purposes of the site, though its uses have been modified. The full visitor's enjoyment must be guaranteed, in relation to the site and its social, cultural and aesthetic context, and at the same time its maintenance and conservation in the long term must be also guaranteed so that its integrity is safe through time.

This balance may be altered by some obstacles that hinder a harmonic relationship between heritage and tourist activity, especially in developing countries having population settlements with heritage value (coastal, mountainous zones, fishing villages, etc.). There may appear a great pressure by the tourist use of those settlements; this process coexists, among other phenomena, with the increasing number of new constructions, including holiday houses uninhabited most part of the year and the public inefficiency for providing economic resources for conservation and rehabilitation activities (Öznögul, 1998). This situation can be observed in many cities in different countries.

The previous paragraph shows the lack of understanding regarding the nature of sites, which are only analysed according to their "picturesque" features or to the way of life characterising them. The massive incorporation of visitors may represent the risk of incorporating foreign patterns which can destroy the cultural heritage to be preserved.

Certain population settlements with heritage value indicate that the building morphology and their characteristic urban fabric respond to complex factors including climate, safety, economy, politics, socio-cultural and religious factors. That net of interdependent factors creates a strong association between population and the built environment. The open spaces help to place the inhabitants in their social space, influencing the way and intensity of the communication between them. Historically (such as the case of many settlements in Saudi Arabia and nearby zones), the circulation systems were designed for enabling the connection between the residential areas and the agricultural tasks, the cult places and the markets. Many paths were designed as labyrinths for disorienting the rivals in case of attack. More recently, the land subdivision according to a grid, the abandonment of the compact fabric for an extended and disorganised urbanisation, surfacing of passages and open spaces for enabling the vehicle access, as well as the excessive tourist interest have destroyed the original balance (Saleh, 1998).

3. Analysis of the main variables

3.1 Built heritage and the economic variable

Including the economic variable into the domain of nature and culture aesthetics has always been suspicious as it seems impossible to value the invaluable. Regarding this, the viewpoints vary with the culture of each people. As Michel Racine says, referring to the "tourism of gardens" in Great Britain and France, "... *garden tourism has become a business, activity considered as the most noble in the mainly protestant countries, when this activity is often suspected in France*" (Racine, 2001).

However, the association between production, commerce and aesthetics has an early background. Significant technological innovations took place not only to improve hunting or recollection efficiency but also to achieve aesthetics goals. In the Aurignacian period (40 thousand to 28 thousand years ago), the Cromagnon man created several techniques for working ivory, including the preparation and use of metallic abrasives for polishing it. They used the ivory for creating beads, earrings and small figures, and seldom for making tools or weapons. Objects made in bones and mammal teeth, fossils, corals, limestone, etc. have been found; the raw material was not chosen at random, and many materials were from geographically far destinations, acquired by means of trade (PNUD, 1998).

Incorporating the economic aspects to this knowledge field has been expanded in the last twenty years, taking into account the following causes, among others:

- Recognising the existence of a "cultural sector" and of "cultural industries", understanding as such, works and practices related with the intellectual and artistic activity, concept that has been broadened, including recreation, sports and free time (Casey et al., 1996).
- The development of specific methods of economic assessment that, starting from the premise that these goods have special features (association of tangible and non-tangible values), take into account aspects that the market disregard. Thus, the economic assessment of natural and cultural heritage allows the existence of a common measure of comparison with other goods, useful when giving priorities to investments, especially in the public sector.

There are different branches of the economy that are directly or indirectly related to the built heritage. This is considered part of the culture economy; thus, it is linked to the theory of property rights (public and private sector, or problems related to different generations), the economy of regulatory instruments (incentives and other measures related to heritage conservation policies), cultural tourism, etc.

A feature of heritage is that this has not been produced intentionally as such, as goods or collection of goods, but it is a heterogeneous whole of elements produced for several purposes and that has come down to us with a deeper meaning than the original one. What today is considered heritage is a social and cultural construction, created and controlled by experts. This construction has been changing throughout time and has social perceptions about what is culturally interesting and valuable. According to Pierre Bordieu, it is about the creation of cultural symbols by "consecration". In the case of heritage, the goals are obtained through standards where, as mentioned before, experts mainly participate (Towse, 2002).

An avoidable consequence is that the heritage stock is always growing, as a result of some mechanisms, such as the designation of goods from certain age or type, the inclusion of new heritage categories, etc. Among other consequences, this situation leads to carry out an economic choice, determining the amount of investments for adding value through conservation or revaluation. As regards its possible tourist or recreational use, either the lack of visitors or the excess of them brings about economic problems to the authorities in charge of heritage.

Once the object and field of study have been defined, it is possible to refer to the different dimensions of the economic research in the domain of culture and heritage (Ost et al., 1998). We are interested especially in two:

- "City or Historic Centre": this dimension is linked to Urban Economy and, more widely, to the sustainable development issue.
- "Tourist": heritage is considered as attractive and its impacts are studied in the local economies.

From the economic point of view, the built heritage is subjected to a double approach. It can be considered as goods, as a product (*commodity*) and at the same time, as service support. Heritage as goods is the monument, the site or building with its physical features. From the point of view of services, it is about the concerned built heritage with the use or function to which it is affected or potentially can be.

3.2 Built heritage and tourism: benefits and risks for sustainability

When including the historic environment built by man, the built heritage comprises a great variety of goods: buildings, old monuments and archaeological sites, designed landscapes and gardens, battle fields, industrial buildings and ruins.

Since the 1980s, the interest for the relationships between cultural heritage and tourism has been strengthened and the economic function of heritage has been explicitly recognised by those responsible for decision-making. Performed studies show that each restored and enhanced site gives rise to a certain amount of jobs in the site and in peripheral activities, generating a cost per visitor that can be estimated (Vincent et al., 1993). If taking into account the annual volume of visitors related with certain attractions (for instance, in France about 7 millions annual visitors for the Eiffel Tower and more than 8 millions for the Louvre), it can be observed that these economic effects, direct and multiplicative, are extremely important.

However, this situation, favourable at first, has negative aspects related with well-identified problems: excessive amount of visitors, the lack of content in cultural sites, the risk of developing a pseudo-cultural tourist offer, etc. The excessive visits threaten the existence of buildings and monuments, along with the progressive loss of its cultural identity. The traditional commerce often changes to *souvenirs* shops. The cohabitation becomes difficult; the original inhabitants often emigrate, leaving the historic centres empty in low season, distorting the local urban characteristics.

The badly-administered cultural offer grows and the projects revaluing cultural heritage with tourist purposes are usually faster than the demand. The success of certain cultural sites attracts tourist, hotel and real estate agents. Thus, some behaviour similar to that destroying several sea or mountain sites is developed around the cultural heritage. Tourism, which at first is an important tool for being aware about the heritage value as basis of local cultural identity, becomes finally a way of trivialising urban and rural landscapes.

As regards heritage in hands of private owners, other problems appear. Many of them think that opening their properties to the public is anti-economic and that visits and economic activities organised for making money damage or destroy the heritage to be preserved. In addition, the fact that the building appears in an official list as protected property may influence negatively in its market value, since future modifications are limited. Buildings or areas to be preserved provide a benefit to the society but create a cost for the proprietors (although there are economic incentives for preservation), who are prevented from altering them in order to obtain other economic benefits (Cassey et al, 1996).

4. Case study and discussion

4.1 The sustainable historic city

An interesting example for the study of the interdependence between the economic, cultural, social and environmental variables is the city and, in particular, the historic city in developing countries. The socio-economic causes that lead to the downfall and destruction of that heritage will be analysed, as well as examples of strategies in order to revert the situation, put into practice in some countries with the help of international organisations.

The protective measure of the urban historic heritage should be taken not only to suitably preserve and revitalise buildings and sites with the aim of satisfying the social, cultural and economic requirements. On the contrary, these actions should be tightly linked to a wider objective concerning the urban identity, considered a requirement for increasing environmental and human quality of the settlements, trying to get rid of degenerative forms of the urban fabric such as decline, insecurity and lack of efficiency. This degeneration results in chaotic and unsafe urban environments, with problems of social and environmental decline. In terms of this, Notarangelo mentions Pierluigi Cervellati, who believes that the memory of a historic city should be assimilated with the memory of human beings. When men lose their memory, they become mad, and the same happens with cities (Cervellati, 1991, as cited by Notarangelo, 1998).

A historic city must be preserved in its nature of cultural heritage and economic resource. We will not analyse the first issue, which was considered for the first time, at international level by the Athens Charter referring to artistic and archaeological heritage (1933) "... *the architectonic values must be preserved in all cases, either isolated buildings or complete urban nuclei... must be preserved when they are the expression of a previous culture or reach a general interest..."*

We will analyse which is the material benefit, beyond the spiritual one, that preservation offers to the process of urban development and management. It is interesting to study which is the economic advantage that the public or private sector can obtain from the conservation of historic centres. In 1977, the Charter of Machu Picchu introduces for the first time the consideration of the material value underlying in the conservation of historic centres, joining the economic value to the cultural one *"... the action of preserving, restoring and recycling the historic environments and architectonic monuments must be integrated in the vital process of urban development, also because it is the only way of financing and managing this operation"* (Machu Picchu Chart, 1977).

In developing countries, the problems present characteristic features. When the residential and economic activities abandon the historic centres, the benefit resulting from the real estate market declines. As the space demand is reduced, so is cash flow, therefore sales decline and become sporadic. Besides, in many cases, the preservation regulations together with the building deterioration increase construction costs, making the restoration little competitive compared with other areas. These trends prevent the private sector from making new investments in the historic centres: investors look for opportunities outside this area, re-feeding the decline process.

The presence of non-productive urban districts or of those districts that do not meet the idea of an efficient and comfortable city leads to a situation which many historic centres experience. They are aside of the development and transformation processes of the contemporary city, and their inhabitants are excluded and forced to live with the phenomena of social and environmental degradation.

The decline of activities also reduces tax collection. This trend, together with the explosive growth of cities, attracts the public investment towards developing zones, accelerating the decline spiral of the historic centres (Rojas, 1999). Some functions in central areas (government, banks, and commercial areas) often continue. However, the construction of other sites in new administrative, university, commercial centres, and residential neighbourhood areas only emphasises even more the economic exclusion of the historic centres and their physical decline.

The deterioration and loss of cultural resources (monuments, building groups, sites of historic, aesthetic, ethnologic or anthropologic value) is due to a great extent to the urban sprawl; to the unplanned development of the urban infrastructure, the inadequate water provision, sewage and pluvial drainage as well as to the lack of maintenance of both buildings and infrastructure.

In many cities, residential, commercial and industrial buildings from the beginning of the twentieth century, with architectonic and historic values have been destroyed or are in danger because of urban development. Due to the high land value, many houses of interest are demolished to give rise to residential or commercial entrepreneurships. Beyond the cultural damage, these facts negatively affect the tourist activity, with the loss of potential benefits for the local communities.

In this situation, it is habitual for an important part of the low income population to live in old buildings in central historic areas, where relatively high socio-economic classes used to live. With the displacement of these inhabitants towards the suburbs, these huge houses were subdivided in order to be used by several families. However, most of these properties neither are connected to the sanitary infrastructure nor receive an adequate system of rubbish collection, that is why their deterioration and downfall rapidly increase.

In this context, strategic or sector plans for reverting this situation have appeared in many cities in Latin America since the 1990s. Rojas thinks that, in developing countries, the activities destined to preservation may undergo three stages. The first is characterised by the pressure that some cultural minorities exert for establishing some control or legislation on this matter. This brings about isolated interventions in specific monuments, generally financed by private philanthropists. Many of these buildings are destined to public use, what leads to a non-sustainable conservation: investments are made now and then due to the lack of systematic maintenance and inappropriate use.

In Latin America, some countries have reached, at least partially, the second stage, in which governments assume responsibilities in conservation. This participation from the estate, potentially positive, brings about other problems: the lack of continuity in the conservation efforts due to budget restrictions and to the volatility of public resources. The Bank for Inter American Development, which gives credits for conservation, has warned that the conservation process, as currently organised and financed, is not sustainable in the long term and represents a heavy burden in the budgets of the public sector that, also has to prioritise the problems regarding poverty in these countries (Rojas, 2001).

The trend should be progressing to a third stage, in which preservation of historic heritage becomes the responsibility of the community as a whole, including the private sector. Sustainability in the long term can be only achieved when the involved social actors jointly collaborate with this aim.

The action of the public sector allows providing the private sector with favourable conditions for its active participation in the process (Rojas, 2001). Firstly, with the provision of stability in the regulatory frame. Investors are always afraid of risks when acting in an area of unknown future. Secondly, showing the feasibility of investments in non-traditional markets is a possible means of deeply encouraging the private investors.

The strategies for modifying this trend are numerous. Firstly, the traditional way of acting is towards actions aimed at the physical improvement of the area (repairs, maintenance, building enhancement, urban infrastructure and facilities. This attitude is not directly aimed at the economic development of the inhabitants, though it may influence it. On the other hand, a plan of socio-economic action may have value for the population, but it does not act on the functional and spatial structure, on the urban shape and its decline. The best way of sustainable recovery of these areas is based on an integrated strategy of interventions, able to solve the demands of the socio-economic development and the conservation of heritage values, without disregarding the spatial identity of the sites.

Some ideas referred to the effective protection of the cultural resources in developing countries are the following:

-   Property ownership: establishing the property ownership and having an updated property registry is a way of promoting the heritage conservation. Legalising land ownership often enables the owners of properties with historic value to obtain credits for improving buildings and this helps the conservation of this kind of property. These strategies must have the conformity of owners, who sometimes do not want to regularise the register of their properties to avoid paying taxes.

- Establishing flexible controls regarding the alternative uses of these buildings to allow an adaptable use and to ensure the conservation. The local authorities can encourage the conservation and restoration of buildings and historic districts, allowing the private sector to adapt old buildings for new uses. This must go together with policies aiming at expanding the economic basis, attracting investments and creating new jobs.

- Register and protection of priority natural resources: it is necessary to identify and register historic buildings which need special protection and thus, determining how this can be carried out in a context of permanent growth or urban development. Sometimes it is not possible or desirable to preserve all buildings. Many proprietors may resist conservation regulations on a private property, unless they are compensated according to the benefits they would obtain from that land if it was liberated for new uses (apartment tower building, for instance). If the list of buildings to be preserved is excessive, it could be impossible for the Estate to face that burden and in the long term it may lead to a deeper process of deterioration in the area (Berstein, 1994).

4.2 Case study: Latin American World Heritage towns

In the context above mentioned, this chapter will refer specifically to sustainability problems caused by the development of tourist activity in urban centres and historic neighbourhoods, particularly in Latin American countries. It is a kind of heritage highly significant and valued by tourists, since it is in the old city neighbourhoods where the distinctive signs of a particular culture can be seen more clearly: the principal architectonic monuments, the public spaces with higher symbolic value and the most significant components of the immaterial heritage.

The case of historic centres in Latin American cities is apt to set an example of the many problems that are related with the triad heritage – tourism – sustainability, especially in its economic and social aspects. With regard to the origin and evolution of Latin American towns, Hardoy (1971) identifies six stages:

a. Pre Columbian period, in which 5% of Latin American territory was occupied by urban cultures. Even if the Americas were totally populated before the arrival of the Europeans, original cultures reached different degrees of development; the most advanced cultures were located in Meso-America (a portion of present Mexico and the Central America) and in the Andean region of South America. At the arrival of the Spaniards, some towns like Mexica-Tenochtitlan or Cusco matched or even surpassed in development and architecture many European cities of the time. In some cases, Spanish towns were settled on the remains of pre-Hispanic ones.

b. Stage of Spanish foundations over the first half of the sixteenth century, based on regional and urban infrastructure of the pre-Columbian cultures.

c. Establishment, by Spaniards and Portuguese of ports, mining towns, forts and reductions. The territorial structure was based on natural resources and on a communication system and the urban basic schemes were defined around 1580 for both Spanish and Portuguese territories. New towns were especially settled along the roads systems linking the production or mining areas with the ports and on the seashore. Spanish towns were settled on the basis of strict legal regulation regarding urban and territorial layouts. The common type was based on a regular grid pattern of streets with a central plaza that constituted the civic, religious and commercial core of the town. Portuguese towns, conversely, were constructed according to more organic urban schemes, sometimes taking into account the topographical features of the setting.

d. Once consolidated institutions and norms of colonial life, there was a period of about two centuries with no significant changes, with an urban scheme that would remain until the arrival of the railroad. During this stage major administrative and trading centres were consolidated.

e. The independent period, which started on different dates depending on the specific countries but was consolidated by the late nineteenth century with the inclusion of countries in the region to the global economic framework, including the massive influx of immigrants in some countries. During this period the railway was incorporated and ports were modernised, the first urban industries and new institutions were settled. Some cities, particularly political capitals and ports initiated a sharp expansion. The establishment of new towns responded to a variety of requirements, among them the consolidation of the boundaries of the new countries, the incorporation of new territories to the productive system, the construction of new ports or the establishment of administrative state or regional capitals. Urban patterns were generally based on the heritage of the previous period.

f. Over the twentieth century, the most significant process was the incorporation of former rural population to urban centres; there was an explosive growth of industrial cities and, to a lesser extent, of provincial capitals, and a decrease of population of rural areas or villages.

Latin American historic centres correspond generally to colonial towns, in a few cases constructed upon the remains of Pre-Columbian cities, which conserved their main urban and architectural features with no major changes throughout the nineteenth and twentieth centuries. In these cases, new developments occurred out of the boundaries of the colonial cities; in other cases instead, extensive renovation over the last two centuries prevented the historic cores from preserving their original features. In the cases where the historic centres preserved their traditional features, these areas generally presented diverse degrees of functional and physical degradation whereas those evolving cities (Sao Paulo, Buenos Aires or Santiago) changed drastically the original appearance. The "discovery" of the historic centres started ain the 1960s and restoration and conservation works have been developed since then. The most prominent Latin American historic centres are inscribed on the UNESCO World Heritage List. In 2007, 38 out of 84 Latin American World Heritage properties were historic towns or centres, a figure that represented 45.23% of the cultural properties and 31.40 % of the total of World Heritage sites in the region.

Latin American historic towns and centres bear some common features if compared with those belonging to other geo-cultural regions; at the same time there are specificities given by their history, urban and architectural features and symbolic content. At the same time, there are particular pressures and constraints. For example, the construction of major development projects, common today in many historic cities around the world, seems not to appear as a real threat, although some examples can be found in the region. However, the strong impact of tourism, social changes, inadequate maintenance of public spaces, buildings and sanitation problems, which are not serious problems elsewhere, tend to appear frequently as specific threats to the integrity, authenticity and sustainability of these towns. A characteristic cycle may be described, as follows:

- The active population abandons the historic centres, with obsolete buildings and infrastructure, moving their economic activities to peripheral areas. The area is then populated by the lowest strata of the real estate market, by unemployed or low-income people.
- Decades later, the public sector, sometimes together with the private sector, decides the rehabilitation of these areas, making them a tourist attraction. If the activity is no properly regulated and the carrying capacity is excessive, there is always the risk of beginning again the decay cycle with new damages to the material and immaterial heritage.
- It may take place an expulsion of local inhabitants due to the high cost that now the land and properties have due to the effect of the new commercial value, giving rise to the social phenomenon known as "gentrification" .

Regarding the impact of tourism on historic towns and centres, it is possible to identify some indicators of what could be considered positive impact on the heritage sites and on the local community. The general idea is that tourism constitutes an opportunity for development; particular indicators are economic benefits produced by tourism, creation of jobs related to tourism, improvement of infrastructure and public spaces and opportunities for education and training. With regard to public spaces and infrastructure, World Heritage towns and urban areas are usually a target for improvement and enhancement. Usually, public spaces are well preserved and maintained; adequate urban furniture and infrastructure are provided or improved. This is an action that contributes to a general amelioration of quality of life of local population and enhances the experience of visitors. The inscription on the World Heritage List and the increase in the number of visitors constitute also an opportunity to restore historic buildings. New uses are often related to tourism. It is quite usual in Latin American towns that old one-family houses, quite difficult to continue with its original use, are dedicated to accommodation for visitors. This can be considered an opportunity in two senses; on the one hand historic buildings are restored and given a new use, generally by private investors, and, on the other , visitors may live the experience of lodging in typical historic houses, which is a means for a deeper contact with the local culture.

With regard to threats caused by unplanned or inadequately managed tourism, it is possible to summarise the situation on the basis of the following scheme (Conti, 2011):

- Replacement of traditional population, gentrification.

A frequent aspect related to management or historic urban areas and its tourist use is the displacing of traditional population or depopulation of historic neighbourhoods. This situation is strongly linked to a process that can be noted equally in developed and developing countries. The inscription of an urban area on the World Heritage List implies generally improvement of public spaces, of services and infrastructure. This takes to a rise in the market values of urban land and buildings and takes almost inevitably to the replacement of population. Traditional inhabitants of many historic areas are sometimes displaced by self decision since they prefer to sell their houses and buy new ones out from heritage areas, something that produces financial benefit. In other cases, they are "pushed" by the pressure of investors or by groups having major revenues who want to buy properties in the prestigious historic areas.

The phenomenon of gentrification means the replacement of typical population of a given urban neighbourhood for another of greater financial resources or more exalted social position. The truth is that this process takes place equally in urban centres around the world, though with higher recurrence in economically disadvantaged countries. The reason that causes this process lies primarily in the fact that the buildings are purchased by individuals or entities, usually affecting them to different uses of the original. Thus, it is common to find old homes converted into hotels, restaurants or shops or even maintaining the residential uses but occupied as secondary homes for short periods throughout the year. This implies that the neighbourhoods gradually lose their population, which means at the same time a crisis regarding some aspects of authenticity. In this case, although the tangible components of buildings may be in good condition, even improved with respect to their previous state, a loss of authenticity of functions appear.

-     Threats against authenticity

With respect to the alteration or distortion of heritage values and message, it is necessary to consider heritage as a set of tangible assets to which values related to history, art or science are assigned. In this sense, heritage is a carrier of meanings that we try to transmit, through the conservation of the material substance, from one generation to another. A proper understanding and interpretation of such values is therefore essential to understand the true meaning of heritage, to ensure proper use and to preserve its authenticity, understood not only as the preservation of the tangible components but also the intangible ones as functions, vocations, associated traditions, etc. In this sense, a conflict often observed with the spread of mass tourism is that heritage becomes a sort of spectacle and object of consumption, without reaching the adequate transmission and understanding of its values.

It may happen that while a heritage site is well preserved and its carrying capacity or limits of change are maintained at appropriate degrees, its dedication to tourism involves risks to its authenticity. This is a situation frequently observed in some heritage categories such as historic centres or urban areas. According to the current theory, authenticity is verified in several ways, taking into account tangible and intangible components; thus authenticity includes the consideration of materials, shape and design, setting, functions and vocations, meanings and traditions associated with specific sites (ICOMOS, 1994).

When we refer to threats to authenticity, we refer not only to damage caused on the material components of heritage but also to the risk on the intangible aspects that influence their authenticity. A typical case consists of many very well preserved historic towns or centres, both buildings and public spaces have good and proper maintenance. But excessive devotion to tourism means that the entire neighbourhood is devoted to visitors; all businesses are dedicated to the tourist, the old residences are now hotels or are destined to gastronomic uses, etc. The problem then is that while the material substance may be well preserved, this area has lost or drastically changed its meaning and its original functions, so there are aspects of authenticity that are really at risk. This aspect tends to be one of the most complicated issues in the management structures of heritage sites.

- Impact on traditional ways of life

Another type of impact is related to social aspects and, especially, to the relationship between the local community and visitors. Particularly when there is an economic asymmetry between them (more specifically tourism in economically disadvantaged regions), it is common that residents consider visitors as an "opportunity" to secure or increase their income. This can include the selling of typical products or the adoption of behaviours that are expected to cause impact on the visitor. It is common to find people who dress or act in a way that is not part of their daily lives; they become a part of the stereotypical image of the site and, therefore, something that the tourist expects to see. The problem is that in this way the resident community, or some of its members, are at the service of visitors' expectations; this implies another manifestation of a threat against the authenticity, in this case referred to lifestyles, habits, behaviours, etc.

Intangible heritage is fundamental not only to determine the outstanding universal value but also the authenticity of World Heritage properties. In spite of problems of depopulation and gentrification, most Latin American World Heritage towns retain a very rich intangible heritage made up by, among other components, music, gastronomy and traditions. This intangible heritage becomes also a tourism attraction and could be jeopardised if adequate safeguarding measures are not defined and implemented. It is worth asking what the limit is to safeguard traditional ways of life or social practices so that they could be preserved as authentic cultural manifestations and not as performances for visitors. What usually happens is that members of local population act in the way visitors expect them to. There is a sort of alienation of the local population in the visitor's expectations; this is one of the most important threats against authenticity.

Within the social aspects some "contradictions" appear between what could be called a positive effect on tourism and reality. When talking about tourism opportunities we have referred to the economic benefit, to the possibility of improving urban spaces and the provision of infrastructure and equipment. Often, the economic benefit is not evenly distributed among the resident community; although there is a general improvement of urban space and infrastructure, some sites are inaccessible to the residents. Many times the cost of access to cultural facilities and entertainment are fixed in terms of tourism, making them inaccessible to local people.

In order to illustrate these aspects, we will introduce two specific cases, the World Heritage properties of Cartagena de Indias, Colombia, and Colonia del Sacramento, Uruguay.

a.   Cartagena de Indias

The historic centre of Cartagena de Indias and its fortresses were inscribed on the World Heritage List in 1984. The property includes the walled city and a set of fortresses located along Cartagena bay. Cartagena was one the most important South-American ports over the Spanish period; the richness of the city and the importance of its port are evident in the architectural monuments (churches, convents and private residences) and in the defence system, since the town was several times attacked by pirates. Not only was the town surrounded by a massive wall but several fortresses were erected along the bay, protecting the entrance to the port. Cartagena is considered the most impressive ensemble of military architecture constructed by the Spaniards in the Americas and, at the same time, one of the most beautiful and well preserved historic centres in Latin America, since much of the traditional urban fabric has been preserved.

It is not strange that the historic centre became a main tourism destination in the region. A joint UNESCO-ICOMOS report of 2006 recognised that *"the historic centre has not undergone substantial physical alterations ... while the use of the urban soil has deeply changed"*. The impact of tourism was the main cause of these changes. Until the 1980s, there were not luxurious hotels in the historic centre; the accommodation offering was limited to hostels of lower-middle level hotels. At the beginning of the twenty-first century, five-star hotels and conferences centres have been installed in former convents, skilfully renovated; palaces and historic houses have been restructured to house charming hotels and hostels and some residences were transformed into second houses for national and foreign tourists. The intense demand has increased the market prices, something that made convenient for residents to sell their properties and to leave the walled city.

Tourism has impacted differently on diverse areas of the historic centre. In the Centro district, the core area of the historic centre where the main institutional buildings are located, some of the positive effects already mentioned above can be noticed, such as the improvement of public spaces, the provision or urban facilities and furniture or new uses for historic buildings (Fig. 2).

Fig. 2. Cartagena, the Centro district. Good state of conservation of buildings and public spaces and high impact of tourism. (Photo A. Conti)

This is the area which exhibits the best state of conservation of the tangible components whereas the intangible ones have changed. The process of gentrification is evident; commercial facilities are related to satisfy the demands from visitors: luxury handicraft, restaurants, bars, night clubs and travel agencies are predominant in this area. San Diego district has traditionally been a more disadvantaged neighbourhood next to the Centro area. The invasive tendency of tourism is more contained, but not less important; it is concentrated around the large hotels and some public squares, and still cohabits, with certain equilibrium, with the traditions of the residual residents. However, this equilibrium is unstable and the increment of the tourist activities could compromise it definitively (Fig. 3).

On the other hand, there are areas where the pressures of tourism are not so evident so far, where traditional local population still lives. The neighbourhood of Getsemaní is the place of residence of low income traditional population. Although the state of conservation of buildings is not as good as in the Centro district and there are some problems with infrastructure, we can still notice the traditional ways of life and uses of the public space. Authenticity is noticeable not only with regard to tangible heritage components but to intangible components as well (Fig. 4).

Fig. 3. Cartagena, San Diego district: balance between traditional life and tourism. (Photo A. Conti)

Fig. 4. Cartagena, Getsemaní district: preservation of traditional population and social life. (Photo A. Conti)

Summarising, Cartagena could be taken as an example of different situations within the boundaries of the historic centre. Economic and environmental sustainability is evident in the districts where the impact of tourism is stronger, while social sustainability is at stake. Conversely, the areas preserving traditional population present deficiencies regarding their state of conservation and quality of life.

b.   Colonia del Sacramento

The historic quarter of Colonia del Sacramento, Uruguay, was inscribed on the World Heritage List in 1995. The origin of the town was a village settled by Portuguese in 1680 on a peninsula by the east embankment of the Plata River, opposite of the then Spanish town of Buenos Aires. The village passed from Portugal to Spain and vice versa several times up to 1778, when it came definitively to Spanish rule. Colonia is an interesting example of merging of different urban and architectural features; although there are not impressive architectural monuments, the historic centre retains much of the typical atmosphere of a colonial town, increased by its setting. ICOMOS recognised that *"the main feature of Sacramento is however, its overall townscape, with its mix of wide main thoroughfares and large squares with smaller cobbled streets and intimate squares. The vertical scale is perfectly preserved, only the church tower and lighthouse rising above the mainly single or two-storeyed buildings"* (ICOMOS, 1994).

Fig. 5. Colonia: good state of public spaces and historic buildings. (Photo A. Conti)

Even before the inscription on the World Heritage List, the historic centre of Colonia had become an important tourist destination, the second one in importance in the country (Assunçao, 2002). It is worth noting that the town is located some two hours by car from Montevideo, the country's capital city, and fifty minutes by ship from Buenos Aires, which comprises some ten million inhabitants within its metropolitan area. The process of gentrification started much before the inscription of the property on the World Heritage List and has continued ever since. The charming atmosphere of the historic centre made that people form Montevideo or Buenos Aires used to buy residences as secondary houses, a process that took to the progressive depopulation of the historic centre and to the rise of prices of land and buildings within the area. In 2002, Uruguayan authorities reported that the price per square metre was more expensive in Colonia than in Punta del Este, the well-known international Uruguayan beach resort. According to Venturini (2008: 11) the population of the historic quarter was some 300 inhabitants in 2008. Public spaces and architectural heritage are very well preserved in Colonia. Over the last forty years, the Honorary Council, the body in charge of the management, and the local government has made significant investment in restoration, conservation and maintenance (Fig. 5).

The impact of tourism on the public space is easily noticeable; some streets have been closed to motor traffic and have become outdoors cafés or restaurants. Historic houses have generally been bought by people who use them as second residences and many buildings have been given new uses such as shops, accommodation facilities, restaurants or cafés (Fig. 6).

Fig. 6. Colonia, the impact of tourism on public space. (Photo A. Conti)

As in other historic centres, whilst the tangible components exhibit a good state of conservation, the authenticity of intangible components is especially at risk because of the impact of tourism. In this case, moreover, the proximity with Buenos Aires results in a one-day excursion is the most typical way of visiting the town, with an average of 3000 visitors per day along the whole year (Assunçao, 2008); most visitors do not use the accommodation facilities and often spend only a few hours in the site.

5. Conclusion

As stated in the Brundtland report, sustainability includes three dimensions: economic, social and environmental. These three aspects were considered by the World Tourism Organization (WTO) to review the definition of sustainable tourism in 2005, stating that an adequate equilibrium should be established among the three. Although indicators to measure sustainable development have been used over the last twenty years and that WTO has been promoting the use of sustainable tourism indicators since the early 1990s, the application of systems of indicators to tourism is more recent and still in a tentative face (Rivas García and Magadán Díaz, 2007). There is no a unique methodological approach for the definition of indicators of sustainable tourism (Blancs Peral et al, 2010); the definition and selection of indicators will depend on specific situations. The WTO has developed a system of core and supplementary indicators; among the former, there are several related to physical and social aspects: social impact (ratio of tourists to locals), developing control (existence of environmental review procedure of formal controls over development of site and use densities), planning process (existence of organized regional plan for tourist destination region), consumer satisfaction (level of satisfaction by visitors), local satisfaction (level of satisfaction by locals) and tourism contribution to local economy (proportion of total economic activity generated by tourism only). Among the supplementary indicators to be used for urban environments it is worth mentioning site degradation, restoration costs, levels of pollutants affecting site and measures of behavior disruptive to site (OMT, 1995).

A sustainable city is that which has the capacity of surviving and adapting to processes of changes, and at the same time, providing an environment quality related to settlement patterns and contexts throughout different times. The urban development needs to pay more attention to issues such as durability, the rational use of energy, pollution, natural and cultural heritage, resource conservation and biodiversity (Marat Mendes, 1998).

Two of the most important aspects for an adequate approach to urban conservation are the commitment and participation by local inhabitants in the process. Heritage conservation must be dealt not only by governments but also by all the population. It is no longer a public initiative but a community project (Bonnette, 2001).

As regards the cities in the developing world, and as consequence of the serious social and environmental problems, conservation of cultural heritage is not often seen as a priority. However, it must be taken into account that the destruction is generally irreversible. Therefore, their value and the information they contain is lost forever (Berstein, 1994).

Finally, it is observed that in both developed and developing countries, the tourist activity is growing, that is why it is urgently needed to implement practical measures destined to achieve the "sustainable area" which was previously mentioned, in order to balance the increase of visitors with their negative impacts on the natural and cultural heritage. If these limits are crossed, the opposite expected effects may be reached: deterioration and destruction of the involved heritage, by non-controlled visits that that *fragile matter*, as Torsello says, is not in conditions to bear (Torsello, 1998).

With regard to the items discussed in this paper, it is clear that built heritage has become a main tourist attraction. It is perceived by visitors as a testimony of the identity and attractiveness of the place and by stakeholders and residents as a source for revenue and for developing the tourism system. The presented study cases allow defining some conclusions regarding the relationships between built heritage and sustainable tourism:

a. In both cases, it is evident that tourism has become a source of revenues and an opportunity for local economy. Nevertheless, it is not evident how these revenues are distributed among local population. Improvement and enhancement of public spaces are enjoyed by both locals and visitors but some commercial, cultural or entertainment facilities are practically inaccessible for local population.

b.   Public investment is mainly oriented to areas or sectors especially destined for visitors rather than for locals, while private investment is focused on projects that ensure revenues.

c.   There is not a necessary relationship between interventions of restoration or enhancement of built heritage, especially historic buildings, and preservation of the authenticity of the sites. The process of gentrification is a sign of loss of authenticity regarding intangible attributes such as traditional functions or social practices. Nevertheless, this situation does not seem to be a problem for visitors, because they feel attracted mainly by the tangible attributes of historic centres rather than for the real life of local populations.

d.   Sustainability based on economic aspects seems to be evident in both cases, since they can be considered successful from a point of view of generating revenues. The good state of conservation of public spaces and historic buildings allows referring to environmental sustainability as well. What seems to be at stake is social sustainability, on account of the situations explained below, i.e. gentrification, difficulties for local population to access to the facilities especially thought for visitors or acceptance by residents of the changes of use of urban land in favour of tourism uses.

These situations take to rethink how the tourism use of built heritage should be planned and implemented in order to ensure sustainability based on the three above mentioned aspects. Llorenç Prats (2003) challenges the idea that heritage plus tourism necessarily implies development; he proposes that the answer to the question should be "it depends". Prats proposes three alternatives: a strict preservation and a non-expensive presentation of heritage; considering human resources as a significant heritage component (good technicians and low budget) and, finally, considering heritage as an integral instrument for local planning, not a simple instrument but the axis for local planning. This integration among heritage goods, human resources and proper planning could be the clue for a successful relationship between built heritage and sustainable tourism.

6. References

Amarilla, B. (2010). Patrimonio cultural construido ¿Valorar lo invaluable? In: *Patrimonio y desarrollo local en el territorio bonaerense: el caso Chascomús*. Ed. by LINTA/CIC, La Plata, pp. 27-35. ISBN 978-987-1227-06-8.

Assunçao, F. (2002). *Estado de conservación de Bienes Específicos del Patrimonio Mundial. Barrio Histórico de Colonia del Sacramento*. Periodic report on the implementation of the World Heritage Convention in Latin America and the Caribbean. Unpublished.

Berstein, J. (1994). *Land use considerations in urban environmental management*. Urban Management Programme, The World Bank, Washington, D.C. ISBN 0-8213-2723-2.

Blancas Peral, F. J. et al (2010). *Indicadores sintéticos de turismo sostenible: una aplicación para los destinos turísticos de Andalucía*. Revista Electrónica de Comunicaciones y Trabajos de ASEPUMA, Vol. 11, pp. 85-118.

Bonnette, M. (2001). Strategies for sustainable urban preservation. In: *Historic cities and sacred sites*. Ed. By I. Serageldin, E. Shlugar et al. The World Bank, Washington, pp. 131-137. ISBN 0-8213-4904-X.

Carta de Machu Picchu (1977). In: *Revista Summa No. 124*, Buenos Aires, 1978, pp. 60-62.

Chichilnisky, G. (1997). *What is sustainable development?* Land Economics Vol. 73, No. 4. University of Wisconsin Press, USA, pp. 467-491. ISSN 0023-7639.

Casey, B., Dunlop, R. & Selwood, S. (1996). *Culture as commodity? The economics of the arts and built heritage in the UK*. Policy Studies Institute, London. ISBN 0-85374-671-0.

Choay, F. (1992). *L'allégorie du patrimoine*. Editions du Seuil, Paris. ISBN 2-02-030023-0.

Conti, A. (2011). The impact of tourism on World Heritage towns in Latin America. In: *Conference Proceedings "World Heritage and tourism: managing for the global and the local"*. Presses de l'Université Laval, Québec, pp. 385-398. Edited on CD. ISBN PDF: 9782763794389

Filion, F. Folley, J. & Jacquemont, A. (1994). The economics of global ecotourism. In: *Protected area economics and policy*. Ed. by Munasinghe & McNeely. World Bank and World Conservation Union, Washington, pp. 235-252. ISBN 0-8213-3132-9.

Hardoy, J. (1972). *Las ciudades en América Latina. Seis ensayos sobre la urbanización contemporánea*. Paidós, Buenos Aires.

ICOMOS, International Council on Monuments and Sites (1994). *The Nara Document on Authenticity*. Available at: http://www.international.icomos.org/ charters/ nara_e.htm (Accessed 17 June 2011).

ICOMOS (1994): *World Heritage List. Sacramento*. Evaluation of the nomination to the World Heritage List. Available at: http://whc.unesco.org/archive/advisory_ body_evaluation/747.pdf (Accessed 17 June 2011).

Lawrence, K. (1994). Sustainable tourism development. In: *Protected area economics and policy*. Ed. by Munasinghe y McNeely. World Bank and World Conservation Union, Washington, pp. 263-272. ISBN 0-8213-3132-9.

Lichfield, L. et al. (1993). *Conservation economics*, ICOMOS, International Scientific Committee, Sri Lanka. ISBN 955-613-043-8.

Meppem, T. & Gill, R. (1998). *Planning for sustainability as a learning concept*. Ecological Economics Vol. 26 No 2, University of Maryland, USA, pp. 121-134. ISSN 0921-8009.

Notarangelo, A. (1998): Recovery of historical centres. In: *Proceedings XXV IAHS World Housing Congress*, Lisboa, Portugal, Volume 1, pp. 156-162. ISBN 972-752-024-3

Morin, E. (1973). *Le paradigme perdu: la nature humaine*. Editions du Seuil, Paris. ISBN 2.02.005343.8

OMT (Organización Mundial del Turismo) (1995): *Lo que todo gestor turístico debe saber. Guía práctica para el desarrollo y uso de indicadores de turismo sostenible*. OMT, Madrid.

Ost, C. & Van Droogenbroeck, N. (1998). *Report on Economics of Conservation. An Appraisal of theories, principles and methods.* ICOMOS International Economics Committee, ICHEC Brussels Business School.

Özgönul, N. (1998). Interrelaciones entre el turismo y el uso de asentamientos tradicionales-históricos. Caso objeto de estudio: una pequeña ciudad en la costa del Egeo de Turquía, Alacati. In: *Libro de Actas del IV Congreso Internacional de Rehabilitación del Patrimonio Arquitectónico y Edificación, La Habana.* CICOP, Tenerife.

Pearce, D. & Mourato, S. (1998): *The economics of cultural heritage.* CSERGE, University College London (report).

PNUD, Programa de las Naciones Unidas para el Desarrollo (1998). *Informe sobre desarrollo humano.* Ediciones Mundi-Prensa, Madrid. ISBN 84-7114-771-8.

Prats, Ll. (1997).*Antropología y patrimonio.* Ariel, Barcelona. ISBN 84-344-2211-5.

Prats, Ll. (2003). *Patrimonio + turismo = ¿desarrollo?* Pasos, Vol. 1, Nº 2. pp. 127-136. Available at: http://pasosonline.org (Accessed 7 July 2011).

Racine, M. (2001). El turismo de jardines en Europa y particularmente en Francia. In: *Patrimonio paisajista: turismo y recreación.* Ed. by LINTA/CIC, La Plata, pp. 65-76. ISBN 987-98485-3-5.

Rivas Garcia, J. & Magadán Díaz, M. (2007). Los Indicadores de Sostenibilidad en el Turismo. In: *Revista de Economía, Sociedad, Turismo y Medio Ambiente,* RESTMA Nº 6, 2007, pp. 27-61.

Rojas, E. (1999). *Old cities, new assets.* Inter American Development Bank, John Hopkins University Press, Washington. ISBN 1-886938-62-8.

Rojas, E. (2001). Revitalization of historic cities with private sector involvement: lessons from Latin America. In: *Historic cities and sacred sites.* The World Bank, Washington. ISBN 0-8213-4904-X.

Salch, M. (1998). *Transformation of the traditional settlements of southwest Saudi Arabia.* Planning Perspectives Vol. 13 No. 2, E & FN Spon, London, pp. 391-400. ISSN 0266-5433.

Torsello, B. (1998). Architectural conservation as an economic resource. In: *Libro de Actas del IV Congreso Internacional de Rehabilitación del Patrimonio Arquitectónico y Edificación, La Habana.* CICOP, Tenerife.

Towse, R. (2002). The cultural economy of heritage. In: *The economics of heritage. A study in the political economy of culture in Sicily,* Edward Elgar, UK, pp. 3-19. ISBN 1-84376 041-X.

UNESCO, World Heritage Committee (2008). *Operational Guidelines for the Implementation of the World Heritage Convention.* Available at: http://whc.unesco.org/archive/opguide08-en.pdf (Accessed 17 Juin 2011).

Venturini, Edgardo (2008). *Quartier Historique de la Ville de Colonia del Sacramento, Uruguay. Rapport de la mission ICOMOS, 9 - 11 juin 2008.* Available at: http://whc.unesco.org/en/list/747/documents/ (Accessed 22 June 2011).

Vincent, J. & Patin, V. (1993). Patrimoine culturel et tourisme en France. In: *International Scientific Symposium "Economics of Conservation" Proceedings.* ICOMOS, Sri Lanka, pp. 116-118. ISBN 955-613-045-4.

Wells, M. (1994). Parks tourism in Nepal: reconciling the social and economic opportunities
    with the ecological threats. In: *Protected area economics and policy*, World Bank and
    World Conservation Union, Washington, pp. 319-331. ISBN 0-8213-3132-9.

Young, K. (1997). *Wildlife conservation in the cultural landscapes of the central Andes*. Landscape
    and Urban Planning, Vol. 38, No. 3 & 4. Elsevier, Amsterdam, pp. 137-147. ISSN
    0169-2046.

7

Croatian Tourism Development Model – Anatomy of an Un/Sustainability

Lidija Petrić
University of Split
Croatia

1. Introduction

Four identifiable development paradigms have, at one time or another, dominated development thinking, i. e. modernization, dependency, neoliberalism and the alternative development paradigm (Southgate & Sharpley, 2002; Sofield, 2003; Sharpley, 2009). The last emerged in response to the apparent failure of mainstream, economic-growth based models to deliver development (Sharpley, 2009). Opposite to the other three it has been focused on the content rather than the form of development. Nerfin (as cited in Sofield, 2003; 63) has specified the following premises constituting the alternative development paradigm:

- It is needs-oriented (being geared to meeting human needs both material and non-material);
- It is endogenous (stemming from the heart of each society, which defines in sovereignty its values and the vision of its future);
- It is self-reliant (that is, each society relies primarily on its own resources, its members' energies and its natural and cultural environment);
- It is ecologically sound (utilizing rationally the resources of the biosphere in full awareness of the potential of local ecosystems as well as the global and local outer limits imposed on present and future generations);
- It is based on self-management and participation in decision-making by all those affected by it, from the rural or urban community to the world as a whole, without which the goals above could not be achieved.

With development being increasingly linked with environmental sustainability, from the late 1980s alternative development effectively became synonymous with sustainable development. However some authors suggest that alternative development model is more focused upon specific societal contexts at specific times while sustainable development adopts a much broader focus, in terms of space and time (it is a global phenomenon and seeks for fair and equitable development for all people both within and between generations) (Sharpley, 2009; 45).

However, despite possible dissimilarities regarding the scope and extent between the notions of alternative and sustainable development, the latter has overwhelmed literature and attracted debate and analysis from virtually all academic standpoints. Many authors have striven (though unsuccessfully) to find a single all-purpose definition of sustainable development. Yet at the time when Steer and Wade-Gery wrote their article (1993, as cited in Sharpley & Telfer, 2002) over 70 different definitions were proposed and today they are probably even more numerous. Although the origins of the concept can be traced to the 1960s and the coincidence of the perceived environmental crisis and a global institutional response the most widely cited definition of the concept is given in the so called Bruntland's report stating that "development is sustainable if the present satisfaction of needs does not question the ability of the future generations to satisfy their needs" (World Commission on Environment and Development's [WCED], 1987; 4).

Till today no universally acceptable practical definition of sustainable development has been adopted. However the intention of this chapter is not to add to the already substantial literature on what are regarded as useful approaches to theoretical concepts of sustainability. It accepts that there are many differing approaches to sustainable development and that different policies and practices may be appropriate in different circumstances (Sharpley, 2009). Its main objective is to investigate, through an analysis of the specific case study, whether tourism development model effective in the Republic of Croatia, a well known tourist destination promoted as the "Mediterranean as it once was", is based on the principles of sustainability. The author has analyzed this model by scanning it from all the three aspects of sustainability, i.e. economic, environmental and the social one. For the purpose of a deeper investigation into this matter a desk research has been conducted consulting a substantial amount of sources, such as books, papers, and research studies of which quite a few are based on questionnaires, strategic documents, newspaper articles as well as web posts. The author of the chapter has participated herself in several studies referenced here.

The chapter is structured as follows: after a brief overview of the three main sustainable development aspects, an explanation is provided of the costs tourism development poses globally and locally. Then the concept of sustainable tourism development is introduced expounding in which way tourism has to be developed if the crucial resources are to be preserved in the long term and benefits equally spread among all the stakeholders. The main part of the study endeavours to draw the three main aspects of sustainability together blending the theoretical issues with the practical experience from the case study. The final section, i. e. conclusion, briefly considers the future mechanisms for managing tourism in Croatia in order to make it more sustainable.

2. The aspects of sustainable development

Although there is no universally acceptable practical definition of sustainable development, the concept has evolved to encompass three major aspects of sustainability: economic, social and environmental (Figure 1).

The *environmental sustainability* focuses on the overall *viability and health of ecological systems*. Natural resource degradation, pollution, and loss of biodiversity are detrimental because they increase vulnerability, undermine system health, and reduce resilience. This aspect of sustainability has been the most often discussed through the literature by numerous authors such as Hall, C. M. & Lew A. A. (1998), Hall, D. (2000), Weaver (2006), and many others.

Social sustainability seeks to reduce vulnerability and maintain the health of social and cultural systems by strengthening *social capital* through *empowerment* (Simmons, 1994; Sofield, 2003; Petrić, 2007; Petrić & Pranić, 2010). Preserving cultural diversity and cultural capital, strengthening social cohesion, partnership and networks of relationships are integral elements of this approach (Munasinghe, 2003).

The *economic sustainability* is geared mainly towards improving human welfare, primarily through *growth* in the consumption of goods and services. Economic *efficiency* plays a key role in ensuring both efficient allocations of resources in production, and efficient consumption choices that maximize utility. Problems arise in the valuation of non-market outputs (especially social and ecological services), while issues like *uncertainty, irreversibility and catastrophic collapse* pose additional difficulties (Pearce & Turner 1990, as cited in Munasinghe, 2003).

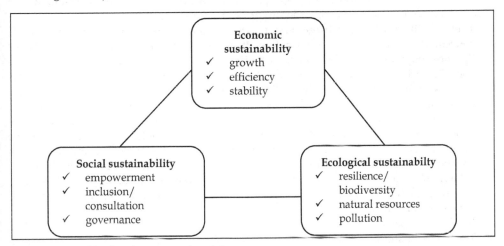

Fig. 1. Sustainable development triangle[1]

3. Could tourism become sustainable?

In parallel with the evolution of sustainable development discourse, concerns about the environmental and social impacts of tourism have escalated in recent years. The main causes for such a rise of concern may be listed as follows:

- International tourist arrivals reached 880 million in 2009 with the projections of reaching 1.5 bill arrivals worldwide until 2020 (United Nations World Tourist Organisation [UNWTO], 2010).
- Such a huge number of people travelling around the world, using the most valuable resources generate not only positive but also a lot of negative effects especially when it comes to the environmental issues. Among them the ones that ought to be specially addressed are:
  - Deterioration of natural resources (fresh water, land and landscape, marine resources, atmosphere and local resources), which may be resilient, but can deteriorate rapidly if impact exceeds tolerable limits (carrying capacities);

---

[1] Adapted from: Munasinghe, M. (2003). Analyzing the nexus of sustainable development and climate change, an overview, Organisation for Economic Co-operation and Development [OECD], 4.03.2011, Available from: http://www.oecd.org/dataoecd/32/54/2510070.pdf

- Disruption of wildlife and habitats, including vegetation, endangered species, use of forest resources, intrusion into fragile areas with sensitive ecosystems;
- Creation of pollution and waste contaminating the land, fresh water sources, marine resources, as well as causing air and noise pollution. Thus, international and domestic tourism emissions from three main sub-sectors (transport, accommodation and activities) are estimated to represent between 3.9% and 6.0% of global emissions in 2005, with approximately 40% of the total being caused by air transport alone (United Nations Environmental Programme [UNEP] & UNWTO, 2008; 33). Environmental problems caused by tourism appear to be even harder since environment, especially the natural one, is a basic resource that tourism industry needs in order to thrive and grow.

- There is still an uneven distribution of tourist flows in terms of their dominant regional concentration. Thus despite the negative effects of the world financial crisis, Europe, as the most popular macro-destination accounted for 52% of international tourist arrivals and 48% of international tourism receipts in 2009 (decrease by 6% in terms of arrivals and 7% in terms of receipts in real terms). As already mentioned, World Tourism Organisation (2010) forecasts that international arrivals are expected to reach over 1.56 billion by the year 2020 out of which 717 million arrivals only in Europe. Of these worldwide arrivals in 2020, 1.18 billion will be intraregional and 377 million will be long-haul travellers. After the tragic happenings of September 11th, securities concern and tighter visa policy over travel to Europe and USA have led to change in the travel behaviour of tourists. This has forced tourists to take their holidays in their own countries, or the region, thereby providing much-needed impetus to regional tourism development but at the same time producing higher pressures on the destinations' natural and social capacities.

- Tourism still shows high seasonal concentration hence posing additional pressure on destinations' capacity to cope with tourists and their numerous activities in relatively short a period of time. In 2009, most of the European residents took holidays in the third quarter of the year, with more than one in three holiday trips made in July, August or September. When taking into account the duration of the trips, the seasonal pattern was even more pronounced, with EU residents spending 46 % of all nights away on holiday in the third quarter of 2009. Short holiday trips, domestic holidays, and business trips tended to smoothen the seasonality of tourism demand. The increasing popularity of short trips slightly reduced the seasonal bias in the period 2004-2009 (Demunter, 2010).

- Tourism is a global phenomenon but locally generated; as such it has to be deeply embedded into a local community. Moreover, local community itself is not only a physical space within which tourism occurs but also a highly complex tourism product. Murphy (1985:153) argues that tourism development "relies on the goodwill and cooperation of local people because they are part of its product." Hence, tourism, being a local community job and strongly affecting community life, requires proactive approaches based on broad participation by stakeholders, which would contribute to more effective policies and plans. This would increase the opportunities to realize the full social and economic potential of the tourism industry.

Due to the elaborated features of tourism industry leading to possible deviations during the process of its development, numerous authors such as Haywood (1988), Bramwell and Lane

(1993), Hall C. M. and Lew (1998), Timothy (1998), Butler (1999), and many others recommend a number of principles that ought to be followed in order to achieve sustainable tourism development. These principles are summarized by Southgate and Sharpley (2002: 243) in the following way:

- The conservation and sustainable use of natural, social and cultural resources is crucial. Therefore, tourism should be planned and managed within environmental limits and with due regard for the long-term appropriate use of natural and human resources. Many studies have been done and implemented so far in order to bring conservation ideals into tourism. Good example is, for instance The World Wildlife Fund for Nature Arctic Tourism Project whose goal was to enable communities, tourists and operators to work together towards a more sustainable tourism (Mason et al., 2000). Many destinations from around the world witness implementation of different hard and/or soft measures aimed at conservation and sustainable use of resources. They include dispersal strategies that 'dilute' tourism related activity and help in distributing employment and revenue benefits more equitably, such as in the case of Maldives, then strategy of spatial and temporal concentration which could contribute to the attainment of sustainable tourism within the destination as a whole (the Gold Coast of Australia illustrates this phenomenon, wherein the vast majority of tourism activity occurs along a narrow coastal strip occupying less then 2 % of the City Council area) (Weaver, 2006). Visitation caps facilitate strategies based on fixed or flexible carrying capacities, depending on whether they apply to absolute numbers or rates of growth. Quotas most commonly used in high order protected areas as well as in a small number of countries are the most formal type of visitation cap and are often used to abet the objectives of zoning system. User fee increases, also commonly employed in protected areas, provide an informal capping effect by reducing the number of potential tourists who can afford to visit the affected site (Weaver, 2006).
- Tourism planning, development, and operation should be integrated into national and local sustainable development strategies. In particular, consideration should be given to different types of tourism development and the ways in which they link with existing land and resource uses and social-cultural factors. A good example of the above is the model of rural tourism which produces multiple benefits for rural population in terms of producing additional income by renting their accommodation, by selling home produced food and drinks and by using local culture as a part of a tourism product. Examples of successful development of rural tourism could be seen everywhere, especially through Europe (Austria, Switzerland, Germany, Belgium, Italy, UK, etc.), particularly in the light of the efforts European Commission has done to this end so far (Veer & Tuunter, 2005; as cited in Petrić, 2006).
- Tourism should support a wide range of local economic activities, taking environmental costs and benefits into account, but it should not be permitted to become an activity which dominates the economic base of an area.
- Local communities should be encouraged and expected to participate in the planning, development and control of tourism with the support of government and the industry. Particular interest should be paid to involvement (empowerment) of indigenous people, women and minority groups to ensure equitable distribution of the benefits of tourism. Good examples of such community involvement in tourism that provides exposure of tourists to local life styles and generates benefits directly to local population might be

the cases of village tourism in Senegal and Sri Lanka (Inskeep, 2006), but can also be found elsewhere, in developed as well as underdeveloped countries. However it is important to stress that in most of the developed countries community consultative arrangements are normative parts of development while in developing countries such a concept may be opposed by the elites running such countries due to the element of power sharing (Tosun, 2000).

-   All organizations and individuals should respect the culture, the economy, and the way of life, the environment and political structures in the destination area.
-   All stakeholders within tourism should be educated about the need to develop more sustainable forms of tourism. This includes staff training and raising awareness, through education and marketing tourism responsibly, or sustainability issues amongst host communities and tourists themselves.
-   All agencies, organizations, businesses and individuals should operate and work together to avoid potential conflict and optimize the benefits to all involved in the development and management of tourism. A number of examples and cases of cooperation among different stakeholders (in Canada, USA, Brazil, Eastern Europe) have been presented in the book "Tourism Collaboration and Partnerships" edited by Bramwell and Lane (2000).
-   In addition, a principle underlying fair distribution of tourism benefits among the members of the local community and internalization of costs produced by tourism stakeholders has to be stressed too.

As seen from the above, sustainability refers to the capacity for continuance of any destination and is, therefore, a function of complex inter-relationships between society and natural resources, a myriad of socioeconomic and political structures and local-scale management decisions[2]. It depends above all on recognition and utilisation of local social and institutional capital (Southgate & Sharpley, 2002: 255-256).

4. Case study: Tourism development in Croatia

After having elaborated the theoretical framework of the notion, principles and aspects of sustainability and the reasons for their implementation into any model of tourism development, there follows the empirical research based on Croatian tourism development model as a specific case study. It must be noted that this case study draws on a more detailed studies and analyses in which the author has participated so far. In line with the study's goal the main research hypothesis has been shaped:

-   Principles of sustainable development, though institutionally recognized are not implemented into Croatian tourism development model in any of the elaborated areas/aspects of sustainability. Hence, it is not only that the achieved results do not correspond to the real abilities but also resources have been seriously endangered by tourism development so far.

---

[2] Very useful source of information regarding sustainable tourism development with a number of case studies and good practices is the "Sustainable Tourism Gateway" web site. It was set up on 27 September 2008 - the World Tourism Day, by The Global Development Research Centre in order to develop awareness and educate on issues related to sustainable tourism, to assist in policy and programme development, and to facilitate monitoring and evaluation. 4.09.2011, Available from: http://www.gdrc.org/uem/eco-tour/st-about.html

In order to prove the above hypothesis, a conceptual model has been introduced. It reveals the main factors featuring the three main aspects of sustainability and links them with the issues that ought to be considered in order to achieve sustainability.

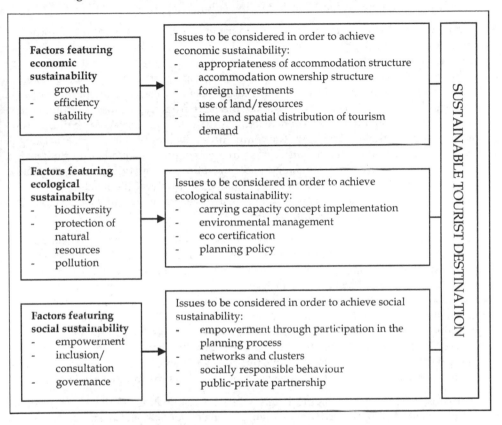

Fig. 2. Conceptual model presenting relations between factors and issues related to different aspects of sustainability

4.1 Croatian tourism - is it economically sustainable at all?

To answer this question it is necessary to reassess the development model of Croatian tourism so far. It started to develop in the late sixties, at a time when the country's competitiveness was largely defined by its inherited, comparative advantages which got aligned with the long-dominant developmental paradigm of the so-called "sun-sea-sand" or mass tourism. Exclusive reliance on natural comparative advantages led to stagnation in the development of higher value-added products. Consequently, the growing influx of tourists did not result in proportionally higher foreign exchange earnings from their spending, and Croatia was on its way to becoming a "low-cost" destination. Today however, it is neither low cost anymore nor necessarily good value for money. According to the results from the latest "TOMAS" research on attitudes and consumption of the foreign tourist in Croatia (Marušić et al., 2010), in comparison with some Mediterranean countries, elements such as entertainment, cultural manifestations, shopping opportunities, sport and recreation facilities, local transportation quality, beach facilities, etc. are still the "Achilles' heel" of the Croatian tourism offer. Hence the average tourist consumption per day is only 58 €, out of which only 7.12 € is spent on sport, recreation and cultural activities.

What caused such a situation? A considerable part of the "blame" lies in the political and macroeconomic circumstances (command economy in the first place) to which the Croatian tourism industry was subject at earlier times. This, however, does not provide a satisfactory explanation for the actual situation. Namely, in the period after gaining its independence (1991), Croatia has unfortunately witnessed merely pale and cosmetic attempts to change the tourism development paradigm, meaning the shift from valorisation of comparative advantages (predominantly based on exquisite natural resources) to building and promoting its competitive advantages. The competitive advantages of a nation, according to Porter (1990), are based on the advanced resource preconditions founded on knowledge, developed infrastructure, high technology and innovation. Competitiveness is therefore a multidimensional phenomenon that must be achieved not only at the level of a firm, but also at the level of an individual, a sector, and a state in general which, in the case of Croatia has still not been recognized to the full extent.

Apart from the misunderstanding of the concept of competitiveness, there has often been a lack of understanding of the true meaning of tourism and its effects and significance for the Croatian national economy. Namely, tourism is frequently discussed as if it were constituted solely by the hotel sector which in turn gives a wholly inaccurate image of the diverse economic effects of the tourist industry. At the same time even the hotel sector operation shows an unfavourable picture. According to the Ministry of Tourism (2010) its share within global accommodation capacities in 2009 accounts for as little as 12.5 % and tourist resorts represent 3.2 % of the total share. Besides, the share of the high-quality hotel and resort facilities is extremely low, i. e. the five star hotels account for 3.9 % while the four star hotels represent 24 % of the total hotel beds; over 54 % are the three star hotels.

Such an unsatisfying picture of the hotels with the domination of the lower category capacities, results in lower occupancy[3], high seasonality of operation[4] and consequently poor performance of the entire sector (Ministry of Tourism, 2010). To prove these statements some results of the hotel sector performance research are presented:

-   In 2009 the average revenue per room (REVPAR) of the Croatian hotels is only 17.6 € as compared to the European and the world's hotels' average REVPAR which is 57.89 € and 54.5 € respectively;
-   Out of 344 hotels analyzed in the study (which is more than 60% of the total), 38.7 % of them are producing loss;

---

[3] Thus the hotel accommodation capacities prove to be used for only 118 days in 2009, or 32.3 % on an annual scale. The use of other forms of accommodation capacities remains even lower; tourist resorts 31.8 %, 16.4 % for camping sites, 11.8 % for private accommodation. The use of hotel accommodation is far below the limits of profitability.

[4] 51.6 per cent of arrivals and 63 per cent of overnights have been realized only in July and August; in the period from May till September over 91 per cent of overnights and 83 % of arrivals have been realized.

- The sector's indebtedness, though already high, has the rising tendency (Horwath Consulting, 2010);
- Croatia is perceived as a country with pricey labour. Hence an average share of the labour costs within the Croatian five and four star hotels' revenue is 5% higher than in the same category hotels of several competitive countries (Spain, Austria, Bulgaria and Monte Negro) (Ivandić et. al., 2010);
- Croatian hotels are on average 45 years old and the period since their last renovations is on average more than 5.7 years (Horwath consulting, 2010);
- Due to the problems Croatian hotel sector is evidently faced with, there are quite a number of hotels that are still to a higher or lesser extent owned by the Croatian state through Croatian Privatization Fund. Namely, according to the latest data presented at the Croatian Privatization Fund's web site, there are 9 hotels in which the state's property share is above 50% and in 44 hotels the state's property share is below 50 %[5].

It is obvious that development problems of the Croatian hotel industry can be solved only by considerable investments aimed at repositioning the entire hotel industry, along with the changes in the development strategy and the completion of market infrastructure in the country. Unfortunately, investments into new assets in the Hotel and Restaurant (HR) sector in the period from 2000 to 2009 make up only 5% of the total investments into Croatian economy while direct foreign investments into the HR sector participate with only 2.5 % in the total direct foreign investments in the country (Croatian Bureau for Statistics, 2010). However, it is to be noted that with the imminent Croatian accession to the EU the investment conditions have been gradually improving and the number of foreign investors in Croatian hotels has gradually been rising reaching total of 107[6] in 2010 (Novak et al., 2011). Among them 65 are owned by companies from the neighbouring countries (Austria, Italy, Hungary, and Bosnia and Herzegovina), which may be explained by geographical proximity, the size of potential market, and cultural similarity. Most of the foreign owned hotels are located in the Istria County (36), then in the Split-Dalmatia County (25), the Primorje-Gorski Kotar County (22), the Zadar County (10), the Dubrovnik-Neretva County (8), the Šibenik-Knin County (3), the City of Zagreb (2), and the Krapina-Zagorje County (1). With the exception of the latter two, all other counties are coastal. Such uneven distribution of hotels results from the fact that most arrivals and room-nights are realized along the Adriatic coast where most foreign-owned hotels are located. Namely, out of 56 million room-nights and 10.6 million guests who visited Croatia in 2010, over 90 % were realized along the coast[7]. Since distribution of demand has been the same since the earlier time of Croatian tourism development, it gives an impression that thus far, increasing number of tourist arrivals, without any spatial or time consideration, is the centrepiece of Croatia's tourism developmental strategy. Attempting to attract as many investors (above all the foreign ones) as possible, with no cost considerations, coastal cities' and municipalities' authorities supported by different profit seeking lobbying groups and individuals have been adjusting spatial plans to the needs of the potential investors in terms of changing the purpose of the land zones from agricultural into the building ones.

---

[5] 26.05. 2011, Available from:
http://www.hfp.hr/UserDocsImages/portfelji/portfelj_hr_100.xls, Croatian Privatization Fund
[6] Foreign hotel is here defined as a hotel present in Croatia by share of ownership above 10%.
[7] 20.06.2011, Available from;
http://www.mint.hr/UserDocsImages/t-promet-2010.pdf, Ministry of tourism

Apart from the unsatisfying structure, quality and territorial distribution of hotels, their high seasonality and unsatisfying business results, high level of the state's intervention in the hotel sector's portfolio, etc. there are also few other problems that deserve attention as they are showing economic inefficiency of the entire tourism industry, not just the hotel sector.

First of all, Croatia has been experiencing the booming development of the households renting accommodation, the so called 'private accommodation'! In 2009, out of 969,726 beds, 435,295 (or 45% of the total) belong to this type of accommodation.[8] According to the state authorities dealing with the tourism inspections, 'private accommodation' accounts for 80 % of the total accommodation capacities on the Croatian coast. There are approximately 52,000 families registered to rent the accommodation, with, according to some estimation, as much as 10 % of that number renting rooms illegally (which brings us to the number of almost one million of unregistered arrivals and six million overnights) (Kljenak, 2011). Due to the fact that Croatia has in recent years faced a complete breakdown of the traditional industrial production which forced people to turn to tourism as an alternative, Croatian authorities have tolerated such huge and still growing private accommodation capacities. By this they have been buying social peace. The problem of private accommodation capacities has yet another dimension, the one referring to their illegal building or adaptation/renovation[9] thus damaging the aesthetic and historical image of the coastline and lowering the quality of the overall supply. How is that possible? On one side the reason lies in the fact that the administrative procedure of getting building permission is long and complicated and hence forces people to choose shortcuts hoping that they will be spared the penalties (due to the poor inspections). On the other hand, local communities highly tolerate illegal building due to the lack of knowledge and awareness of the environmental consequences of such deeds. Furthermore, as an answer to the earlier regime suppressions people strongly support all the profit oriented activities regardless of the costs. Although this situation is gradually changing as Croatia is adjusting its legal framework to the European Union's laws and behaviour standards, the government still does not have a strategy regarding the expected growth of this kind of accommodation, its spatial distribution, the level of desired quality, its relations to the other types of accommodation, etc.

Another important aspect of economic sustainability is the way Croatia is treating its resources used by tourism, especially the public ones. Although this has partially been explained by the example of illegal building of the houses for rental purposes, the following example gives us another perspective into the problem. Namely, the Croatian government has passed the Law on Golf Courses (Narodne novine, No 152/08) thus giving them status of the strategically most important tourist projects. By this Law a potential golf investor can get as much as 30% of the total surface of public land directly negotiating with the local authorities without bidding. Besides, such investor can obtain additional 20 % of the land surface by means of land expropriation from the owners. At the same time agricultural land owned by the state, in spite of being declared country's strategic resource by the Law on

---

[8] This is 40% more than in the pre-war period.
[9] According to the Croatian Association of Urban Planners there are over 150,000 houses built illegally in Croatia; Perica, S. (2011) U Hrvatskoj 150.000 bespravnih objekata, 26.06.2011, Available from: http://www.vecernji.hr/vijesti/u-hrvatskoj-150-000-bespravnih-objekata-clanak-295867

Agricultural Land (Narodne novine, No. 125/08), can be expropriated and turned into a golf course with no compensation for the change of its purpose. Thus the Law on Golf Courses is in contradiction with the Law on Agricultural Land as it prioritises golf course projects over food production, i. e. over agricultural land which is the most important country's resource. As the counties' spatial plans provide for 89 golf courses[10] throughout Croatia, this has provoked numerous debates over the necessity of building so many of them, having in mind potential water pollution and biodiversity threats especially on ecologically sensitive islands and coastal zones consisting of porous limestone. At the same time critics suggest that these laws could encourage land speculators whose only intention is to build as many apartments and villas as possible with the purpose of selling them on the real estate market instead of developing tourist resorts that might work throughout the year and employ local people. Such a scenario related to the golf course project on the top of the hill above the city of Dubrovnik has been recently disclosed becoming an object of intense public disputes (Šutalo, 2009).

Finally, to conclude the session dealing with the economic (un)sustainability of Croatian tourism, the review on its financial results has to be given. In the financial year 2010, revenues from tourism reached € 6.24 billion[11], representing a slight decrease of 1% over the previous year, but a decrease of 8.4% over 2008 (Hrvatska Narodna Banka, 2010). Even such a relatively small amount of tourism receipts as compared to some other European countries of similar size such as Austria, Denmark, Greece, Netherlands, Portugal, etc. (UNWTO, 2011), makes up an extremely important contribution to the Croatian economy, as these receipts are used to cover 55-70 per cent of the foreign trade deficit in the recent years (Blažević, 2007). Another significant indicator of the importance of tourism for the Croatian national economy is its impact on GDP. According to the Ministry of Tourism[12] the share of direct tourism receipts in the country's GDP ranged from 19.4 % in 2005 to 14 % in 2010 [13]. However, such a high share of tourism receipts in the national GDP, especially in the earlier years indicates high dependency of Croatian economy on tourism which is in collision with the basic principles of economic sustainability. At the same time tourism leakages are not negligible. Thus, the Tourism Satellite Account for 2006 (World Travel and Tourism Council, as cited in Petrić, 2006) estimated that direct leakages of the overall tourist economy for Croatia account for about $4.5 billion or 36.5 % of its total GDP. The older analysis carried out by Jurčić (2000) who, lacking the updated intersectoral tables of the Croatian economy, adjusted those from 1987 and estimated that in 2000 the share of total import content in tourism economy amounted to 32 % of its GDP. Evidently, the situation with the tourism leakages in Croatia is getting even worse.

---

[10] 26.06.2011, Available from: http://www.business.hr/hr/Naslovnica/Politika/VIDEO-Hrvatska-mala-zemlja-za-velika-golf-igralista

[11] 26.06.2011, Available from: http://limun.hr/main.aspx?id=694037

[12] 26.06.2011, Available from: http://www.mint.hr/default.aspx?id=5778

[13] On the other hand Tourism Sattelite Account for Croatia (World Travel and Tourism Council; 2011) estimates that direct contribution of Croatian tourism to its GDP in 2010 is 11% and total contribution is 26.3 %. Retrieved from: http://www.wttc.org/eng/Tourism_Research/Economic_Data_Search_Tool/ (24.06.2011). Such diverse information on the effects of tourism on Croatian economy (due to the inadequate statistics) is also a proof that so called "strategic sector of economy" is often misunderstood and its real effects are never estimated precisely.

The above analysis of the economic performance of the Croatian tourism industry has shown that the main requirements of sustainability, i. e. *growth, efficiency and stability,* expressed and elaborated through issues such as appropriateness of accommodation quality and quantity, the role of government in the hotel sector ownership structure, attractiveness of tourism industry to foreign investments, the way of using resources for the purpose of tourism development, time and spatial distribution of tourism demand etc., have not been fulfilled in a satisfactory manner. From what has been shown it could be concluded that Croatian tourism has been developing with hardly any strategy. Measures have been put into operation with no respect to the wider context of development and consequences of the bad decisions have never been penalized.

4.2 What about ecological sustainability?

Many of the issues related to the economic aspects of the Croatian tourism sustainability discussed so far are closely connected to its ecological sustainability, these two being the two sides of the same coin. This is especially true when it comes to the mater of resources and land use. Spatial and seasonal concentration of tourists and tourism facilities, illegal building on the coastal zone, land misuse and speculations, are notably ecological problems but they also create considerable environmental costs and in the long term reduce potential economic benefits.

Concentration of too many tourists in a short period of season (from June to September) creates problems with water and electricity supply (especially on islands), different types of pollution (water, land and air pollution) and the consequent change or loss of biodiversity, damage on cultural heritage etc. (Petrić, 2005). Most of the Croatian coast is seasonally highly saturated by tourists and their activities meaning that carrying capacities[14] of the space are not respected thus leading to environmental, socio-cultural and economic changes and the loss of a destination's attractiveness. As an illustration the case of the island of Hvar, one of the most popular tourist destinations, may be used. Hvar is one of the 66 inhabited islands and occupies an area of 299.66 km². According to the 2001 Census (Croatian Bureau of Statistics, 2001), the whole of the island had 11,103 inhabitants with population density of only 37 inhabitants per km². Although the results of the 2011 Census have not been published yet, the trend of depopulation was evidenced a long time ago. In the season of 2010 the number of tourists who visited island was 172,554 realizing 1,132,982 overnights. From these numbers it is easy to count some tourist density indicators, such as:

-   The number of tourists per km² = 575.83 (as compared to 37 inhabitants per km²);
-   The number of tourist overnights per km² = 3.780;
-   The number of tourists per inhabitant = 15.54;
-   The number of tourist overnights per inhabitant = 102.04.

When these indicators are counted at the level of a single settlement, such as the popular town of Hvar, one can get the idea on the level of saturation such island destinations suffer from. Namely the town of Hvar covers only 75.35 km², including the town itself and five

---

[14] Carrying capacity refers to the number of individuals who can be supported in a given area within natural resource limits, and without degrading the natural, social, cultural and economic environment for present and future generations.

settlements in the hinterland. It has only 4,138 inhabitants and realizes 86,216 tourist arrivals and 439,909 overnights, meaning that the number of tourists per one inhabitant is 20.8, the number of tourist overnights per inhabitant is 106.3, the number of tourists per km^2 is 1,144.2, etc. There has been an attempt to count carrying capacities of the city of Hvar, resulting in a proposal of an eco charge introduction at the city level (Taylor et al., 2005). Unfortunately the proposal has failed due to the strong opposition of the local stakeholders, predominantly people from tourism business who thought that this would push up the prices and consequently reduce demand.

Even before this case, there was a Carrying Capacity Assessment Study done for the island of Vis in early 1990-ties (Dragičević et al., 1997), but although the study had been completed, none of the measures suggested by it has been implemented so far.

The problems arising from the overuse or wrong use of resources in the process of tourism development in Croatia can be seen almost everywhere in its coastal area and in all the types of tourism. However, apart from the residential tourism, nautical tourism has been producing most of the ecological problems so far. Why is that so? Adriatic is a shallow sea rich with different endemic species of sea flora and fauna as compared to the rest of the Mediterranean Sea to which it belongs. They are threatened by an ever rising number of sea vessels that destroy their habitats by draining ballast waters and importing invasive species such as algae Caulerpa taxifolia and Caulerpa racemosa that have already invaded those parts of the Adriatic that are attractive to nautical tourists (such as the Kornati Archipelago National Park, the bay of Stari Grad on the island of Hvar, the surroundings of the Mljet Island National Park, etc.) (Fredotović et al., 2003; Petrić, 2003; Petrić et al., 2004; Petrić, 2005). Besides, for the purpose of nautical tourism development, new marinas are being constantly built, rapidly changing the coastal landscape and threatening biodiversity. As for an illustration, the coastal counties' spatial plans provide for 300 new locations aimed at building new marinas with 33,655 new berths (out of which 25,755 in the sea). Hence, together with the existing ones the total number of berths will be 54,675 (Ministry of Sea, Transport and Infrastructure & Ministry of Tourism, 2008). Apart from this, due to the poor control, there are an enormous number of yachts dropping their anchors illegally in hundreds of wild coves scattered along the coast. Illegal anchoring causes not only the loss of economic benefits in terms of unpaid port charges but also produces environmental costs that are to be paid by society and not the polluter. Similar situation is also with big cruisers that pay daily visits to Croatian ports such as Dubrovnik, Split and Zadar. These ports mostly do not have enough capacities to host so many cruisers in terms of inconvenient infrastructure, insufficient system of monitoring and insufficient material and human capacities to cope with possible pollutions. Not less important to mention is that too many cruise tourists visiting destinations like Dubrovnik may cause discontent of the tourists who reside in the city hotels. According to the results of a research on cruising tourism in Croatia (Horak et. al., 2007), there were almost 600,000 cruise passengers who visited Dubrovnik in 2006 (82% of the total number of cruise passengers in Croatia). In the peak days more than 19,000 cruise tourists happen to visit the old city at the same moment, which together with the numerous excursionists, residential tourists and local population poses tremendous pressure on the city's carrying capacities. The study reveals that almost every fourth tourist (23% of the interviewed) residing in the city hotels thinks that such a huge number of cruise tourists affect negatively the attractiveness of the city.

Environmental problems are caused not only by huge number of tourists but also by tourist enterprises and organisations which intentionally or unintentionally (due to negligence) damage the environment. Unfortunately, although the ecological awareness among them has an ever rising trend, implementation of the concepts of environmental management and eco certification in the tourism business sector is still a rare case. Why is that so? First, it is to be noted that no Croatian law, regardless of their number and variety, deals in particular with the issue of resource usage in the tourist sector (Petrić & Pranić, 2009). The issue is defined in a number of environmental and industry laws. Environmental laws deal with the usage of environmental factors such as water, soil, sea, etc. Industry laws, unlike the environmental laws directed to the general issues, regulate the treatment of concrete natural resources in particular industries (such as agriculture, fishing, etc.). The operation of the tourist industry (and thus also of the hotel sector) is based on various natural and cultural resources and therefore it has to comply with the basic principles of environment protection declared by these laws, and particularly by the Law on Environment Protection (Narodne Novine No. 82/1994; 110/2007). According to this law (art. 150-158) all legal entities (including hotels), are liable for the damage incurred by pollution if caused by their operation or negligence. In a hotel this can be for instance emission of oil or excrements into water, emission of gas into atmosphere, dispersion of asbestos dust, etc. In such cases the hotel not only settles its own damage but also covers all the costs caused by measures taken to eliminate pollution (internalization of external costs). However it is not the hotel sector causing such pollutions so often but rather illegally built private accommodation that is leaning on poor communal infrastructure. Poor control of their behaviour is another reason why they easily transfer the environmental costs they produce to the society.

As for the measures potentially stimulating implementation of ecological initiatives and general environmental policy in companies (including hotels), the Law provides the possibility of regulating benefits, tax incentives, and exemption of tariffs for those entities that use less detrimental production procedures (for example use of alternative energy resources, use of environment friendly equipment and appliances) and those that organize disposal of used appliances or their parts, used products and their packaging or use other ways to reduce negative effects on the environment (Narodne Novine, No. 82/1994; 110/2007).

However, due to the already elaborated circumstances Croatian hotel companies are coping with, most of them are unwilling to implement ecological initiatives and general environmental policy in companies, especially through formal systems of environmental management, justifying their reasoning by high initial costs. The exceptions are the hotels operating within international chains whose ecologically oriented operation is the basic element of their competitive strategy. A few hotels in Croatia implement informal measures of environmental management directed primarily to rationalization of energy and water consumption. To promote necessity of acting in an environmentally friendly way, the Croatian Association of Small and Family Hotels provides training for its members in implementation of the environmental management measures and strives to establish environmental quality mark to be awarded to its members. It also collaborates with Croatian Centre for Clean Production that already in 2006 started a pilot project on possibilities of savings in Croatian hotels by implementation of environmental measures.

However, despite these efforts, the survey of the Croatian hotel sector run in 2009 (Petrić & Pranić) showed that only a third (33.3%) of the hotels in the sample[15] had a written environmental policy, despite environment being Croatia's first and foremost tourism 'attraction' (Marušić et al., 2008, 2010). Moreover, given the implied underlying role of environmental protection in Croatia's official tourism slogan (i.e., "Croatia – The Mediterranean as It Once was"), it is interesting that the reported figure in this research is so low. While interesting, this finding does not come as a surprise as the Croatian lodging sector is still hampered with numerous viability issues – i.e. incomplete and/or poorly executed hotel privatization process, unresolved land ownership disputes, and pronouncedly high seasonality. Under these circumstances, it appears reasonable that the adoption and implementation of environmental standards by Croatian hoteliers is still at an early stage.

As for the eco certificates, except for the EU Blue Flag for beaches and marinas, Croatia has not been included in any international eco certification programme. According to the report by the nongovernmental organisation "Lijepa naša" for 2011 there were 116 beaches and 19 marinas with the Blue Flag certificate in Croatia.[16] Despite the seemingly huge number of certified beaches, one has to remember that the Croatian state has a 1,778 km long coast and 4,057 km of the total coastal line and evidently thousands of beaches.

Apart from the Blue Flag eco certificate, some other instruments and tools (institutional, economic and/or management) aimed at implementation of the environmentally friendly behaviour have also been used but mostly sporadically. Thus, except for the zoning which is an institutional instrument commonly used in the spatial plans (Inskeep, 1991), instruments such as eco taxes, environmental management charge (EMC)[17], visitor payback[18], target marketing aimed at attracting visitors of a certain type and in a certain period of year[19], demarketing[20], price policy aimed at tourist demand attraction or reduction, group size limitations, etc. are sporadically used or not used at all in most of the Croatian tourist destinations for the purpose of resolving problems of resource overuse.

---

[15] The Croatia's Ministry of Tourism (MINT) list of 671 officially licensed and categorized facilities under the group HOTELS (hotels [562], apart hotels [11], tourist resorts [46] and tourist apartments [52]) in Croatia for January 2009 served as the sampling frame for this study. The actual study sample consisted of 310 facilities (46% of the sampling frame) belonging to the group HOTELS (210 hotels, 11 apart hotels, 46 tourist resorts and 52 tourist apartments). The 210 hotels in the sample were randomly selected among 562 hotels using Research Randomizer.

[16] 16.06.2011, Available from: http://www.lijepa-nasa.hr/images/datoteke/popis_pz_2011.pdf

[17] The environmental management charge (EMC) is an amount charged to visitors who visit protected areas or some exquisite locations and perform certain tourist activities.

[18] Visitor Payback is the process of asking visitors to a destination to voluntarily support management and conservation of the area, by donating a 'nominal' sum towards its upkeep.

[19] A target market is a group of customers that the business has decided to aim its marketing efforts and ultimately its merchandise. A well-defined target market is the first element of a marketing strategy. Once these distinct customers have been defined, a marketing mix strategy of product, distribution, promotion and price can be built by the business to satisfy the target market.

[20] Demarketing is a little known concept which aims at dissuading customers from consuming or buying some things either because it is harmful or simply because the demand is more than the supply, especially in case of tourist demand. This could be on a temporary or permanent basis.

To conclude: Croatian tourism is evidently not ecologically sustainable as often being declared. Though institutionally recognized, environmentally friendly behaviour of tourism stakeholders has not been widely adopted yet, which can be proved by the poor implementation of the carrying capacity assessment technique, environmental management concept, eco certification programmes, etc. Monitoring of the spatial plans implementation is rarely done consequently leading to voluntarism in the use of land and resources.

4.3 Croatian tourism social (un)sustainability – the cause or the consequence?

Finally, the third issue this chapter deals with is the one referring to the *social* aspect of Croatian tourism *sustainability*. Though some of the issues featuring social sustainability have already been touched to a certain extent, there is a need to get deeper into this area hoping that this would help us understanding reasons of failure in achieving sustainability in the other two areas. To achieve social sustainability is to empower community and its members to get involved in the process of decision making and planning tourism development. The notion of empowerment has entered literature as a generic term denoting a capacity by individuals or a group to determine their own affairs. Recently it has been used across a wide range of disciplines. The issue of empowerment in the non-management literature has largely been centred on women, minorities, education, and politics and viewed from the perspective of powerlessness and oppression. Simmons and Parsons have a summary definition of empowerment as „the process of enabling persons to master their environment and achieve self-determination through individual, interpersonal change, or change of social structures affecting the life and behaviour of an individual (as cited in Sofield, 2003; 81).

When located within the discourse of community development, it is connected to concepts of self-help, equity, cooperation, participation and networking. These concepts, particularly participation in the process of decision making, is a vital part of empowerment since it makes people more confident, strengthens their self-esteem, widens their knowledge and enables them to develop new skills. Murphy (1985:153) argues that tourism "relies on the goodwill and cooperation of local people because they are part of its product. Where development and planning does not fit in with local aspirations and capacity, resistance and hostility can...destroy the industry's potential altogether".

There are four "types" of empowerment, i. e. economic, psychological, social and political (according to Scheyvens 1999, as cited in Timothy, 2003; 152). Economic empowerment is important because it allows residents and entire communities to benefit financially from tourism. Psychological empowerment contributes to developing self-esteem and pride in local cultures and traditional knowledge. Social empowerment helps maintain a community's social equilibrium and has the power to lead to cooperation and networking. Political empowerment includes representational democracy wherein residents can voice opinions and raise concerns about development initiatives (Timothy, 2003).

To what extent should the community and its members be empowered, or how much empowerment would they experience depends on the level of the social capital development in the country and the community itself (Petrić, 2007). Social capital as a set of formal rules/institutions and informal norms of behaviour creates environment in which the process of empowerment is performed. Grootaert and Bastealer (as cited in Vehovec, 2002; 36) speak on three dimensions/levels of social capital, referring to micro, mezzo and macro levels.

Micro level refers to the networks of individuals and households that create positive externalities for the local community. *Mezzo level* is created by associations and networks. *Macro level* refers to social and political environment that shapes social structure and enables development of the norms of behaviour (laws and regulations).

Croatia generally speaking shows a low level of social capital development on all the three levels, which obstructs communities and their members to be fully empowered to master their future in the sustainable manner. According to Hall D. (2000; 449), such a situation in all the post communist countries such as Croatia, could be explained by considering the following issues:

-   The legacy of almost half a century of centralised, top-down civil administration, affording local people little real opportunity to participate in meaningful local decision-making;
-   The often pejorative equating of any form of collective action with the collectivised organisation of communist days; and
-   The well recognised ambivalence of community as a concept, embracing notions of spatial contiguity, social cohesion and interaction, reflexivity, overlain with often misplaced assumptions of shared aspirations and values.

Apart from the legacies of the communist regime, Stubbs (2007) numbered some other interrelated *macro level* factors constraining 'progressive' community development and empowerment in contemporary Croatia:

-   War consequences (physical destruction, mass population displacement, authoritarian nationalism);
-   Economic and social crises and transition causing widening, regional gaps between the affluent, largely urban areas and many of the war-affected areas, now designated as 'areas of special state concern' marked by high unemployment, low human capital, an ageing population, and tensions between settler, returnee, and domicile groups;
-   A strong impact of rapid urbanization, de-industrialisation and the shifting fortunes of tourism which consequently has never succeeded to get embedded within the local population and culture;
-   The proliferation of numerous local government units (127 cities and 429 municipalities) causing appearance of many municipalities, understaffed and unable to raise revenues locally to be sustainable, meaning that decentralisation is increasingly spoken of rhetorically but rarely pursued in practice;
-   Above all, perhaps the most important constraint on 'progressive' community development and empowerment in contemporary Croatia is not so much 'the new social stratification of Croatian society, accompanied by a significant redistribution of social wealth, social power and social esteem', as the deeper meta-level crisis in values and trust which can be seen as both a cause and effect of this redistribution (Malenica, 2003; as cited in Stubbs, 2007). To prove this statement the following information seems to be very convincing: namely, according to the Transparency International Corruption Perception Index 2010[21], Croatia's rating is 4.1 (0 meaning full corruption, 10 meaning no corruption). Out of 178 countries included in this year's index, Croatia ranks 62nd.

---

[21] 21.06.2011, Available from:
http://www.transparency.org/policy_research/surveys_indices/cpi/2010/results

In such a developmental context sustainability principles in tourism development (and development in general) at community level have never been really embedded. Some recent multidisciplinary researches on sustainable development on Croatian coast with the special stress on tourism development issues (Fredotović et al., 2003; Petrić, 2003; Petrić et al. 2004; Petrić, 2005; Vukonić, 2005; Petrić, 2007; Petrić, 2008; Petrić & Pranić, 2010), have shown the following:

- There is an enormous number of agencies, institutes, committees, and such like, all charged with developing and overseeing strategies and programmes in different areas with overlapping, competing and multiple mandates, thus causing difficulties to small understaffed and underfinanced communities to choose the right strategic direction.
- Environmental policies are usually not reflected enough in most of the tourism sector strategies, plans and programmes. There are no institutional, economic or management tools to implement environmentally friendly behaviour (as explained earlier in the text).
- Plans are technically competent, but often unrealistic and not responding to the local needs. The public is included in the planning process post festum and therefore has no faith in plans and does not make an effort to influence them. On the other hand, efforts to involve the public, if there are any, have most usually been ineffective. The key reason is the way that information is presented, largely in a technical and inaccessible form. Hence, although there is a policy to account for public interest and participation, no real attempts are made to achieve it (Fredotović et al., 2003).
- Specifically with regards to biodiversity protection and conservation, local inhabitants and/or enterprises do not recognize how they may gain from it. Protected areas are designed and managed to respond to national and international needs, not local concerns. The value of biodiversity, to the present and future generations, is not well or not properly understood. There is little faith that the benefits of conservation will flow to locals (Petrić, 2008). These findings correspond to the Hall's statement (2000; 449) that in post-communist countries "any ecologically inspired restriction of personal freedom, such as exclusion from environmentally sensitive areas or the banning of such pursuits as hunting, may be seen to echo the half-century of post-war communist imposition, and thereby meet resistance".
- As already mentioned local communities highly tolerate illegal building of houses/secondary residences or other types of construction though positive results of the Ministry for environmental protection, physical planning and construction most recent activities have reduced such behaviour to a certain extent. Unfortunately, this has not been the result of the rise of the ecological awareness within communities but more of the penalties imposed from above;
- A lack of local involvement in tourism development and decision making has also caused local culture being insufficiently valued as a resource for tourist products (Tomljenović et. al, 2003);
- Generally speaking, though there is a commitment of the Croatian government to the principles of Agenda 21 (1992), explicit institutional response to the needs of Agenda at local and regional levels appeared not to be sufficient in the case of Croatia (Petrić, 2007).

In an attempt to counterbalance governmental (macro level) shortcomings, there has been an enormous growth of the number of Non Governmental Organisations (NGOs) in the country. In 2002 Croatia had over 20,000 registered associations of citizens, with 18,000 of these registered at the local level, but only between 1,000 and 1,500 active ones (excluding sport clubs and cultural associations). Many of the NGOs in Croatia have seen multi-sectoral working as a panacea for many of the problems of Croatian society. The 'List of the non-governmental organisations', published by the Ministry of environmental protection, physical planning and construction in 2004 (as cited in Petrić, 2007) speak of 268 NGOs dealing with environmental issues, while the most recent data mention 710 registered environmental NGOs[22]. Most of these are focused on pure ecological problems while a few, such as ODRAZ[23], a Zagreb-based NGO, are focused on sustainable development of communities in Croatia. ODRAZ is, among other things, strongly committed to the revitalization of the Croatian islands through cross-sector cooperation, including local community organizations, entrepreneurs, and tourist associations.

In Croatia there is no legislative obligation for the cooperation of governmental and non-governmental organizations or for the participation of NGOs in decision-making. However with the imminent accession to the EU, Croatia is obliged to adopt the European Community acquis and a common practice proposing consultation with the NGO's in the process of development and decision making. In recent years an interesting trend of growth has been noticed of what have been termed 'meta-NGOs', whose primary purpose is to provide information and assistance to other NGOs. Hence these larger, more successful, but increasingly bureaucratised or meta-NGOs growingly suppress emerging, under-funded, localised initiatives which ought to be true sources of 'social energy' in Croatia, alongside informal community leaders and local activists (Stubbs, 2007).

Generally speaking, Croatia is gradually making progress when it comes to the civil society development. In 2006 National strategy and action plan for civil society development were adopted thus creating preconditions for the more efficient development of civil institutions at *micro level*.

When it comes to the activities oriented towards empowerment of the key stakeholders at the *mezzo level* through strengthening formal and informal *networks*, there are few examples of long-term, consistent, multi-sectoral partnerships for community development, between local governments, associations and NGOs, and particularly businesses (Franičević & Bartlet, 2001; Petrić & Mrnjavac, 2003; Pivčević & Petrić, 2011). They are usually formed at the national, not regional or a local level. It is mostly vertical type of networks that include different business entities whose aim is better use of resources or better placement of their products or services (good examples are the National Association of Small and Family Hotels and Split-Dalmatian County Association of Hotels, both of which gather hotels as well as tour operators, national air company, suppliers and other subjects creating tourism supply chain. Creation of different types of partnership and/or networks of the firms (horizontal and vertical ones) at a regional/community level, that Croatia still lacks, could help in developing and imposing service standards that will raise the competitiveness of the network and destination tourism brand. Such tourism partnerships and networks can substantially improve tourism business performance by transforming their sporadically scattered products into a one-stop-shop selling a wide variety of functionally interrelated tourism products (Mansfeld, 2002).

---

[22] 24.06.2011, Available from: http://www.mzopu.hr/doc/Popis_nevladinih_udruga.pdf
[23] 24.06.2011, Available from: http://www.odraz.hr/hr/home

Apart from networks another type of partnerships that could be nourished at the local level is through *clusters*. "A cluster is a geographically proximate group of companies and associated institutions in a particular field, linked by commonalities and complementarities" (Porter 1998; 78). Unlike networks, clusters have an open membership, they are based on local values such as trust, empathy, cooperation and have a common vision. Operators within local (tourism) clusters can increase their collective markets and capacities by working together. Working through clusters can benefit all parties involved in terms of increased opportunities and revenues. However, many local tourist communities/destinations lack a system dimension and do not have shared vision or common goals. "And destinations that share little more than joint marketing can not be regarded as clusters" (Nordin, 2003; 18). This statement is proved to be true in the case of Croatian local communities oriented to tourism. Except for Istria, a south-western part of Croatia and Zagorje-Krapina county in the north-western part of the country, where some rudimentary efforts in tourism clustering arise, no other tourism region or a community shows any effort whatsoever to this matter.

The concept of the *socially responsible behaviour* of the firms, although being recognized elsewhere, in the tourism industry is still quite unfamiliar. Thus, a report on Corporate Social Responsibility for 2004 points out a number of positive examples of growing corporate social responsibility and business - NGO collaboration (Bagić et al., 2004, as cited in Petrić, 2007), but no examples from tourism industry were evidenced in either this report or the one made in 2007 (Škrabalo et al., 2007). However, an analysis of the particular web sites shows that the concept has been gradually adopted and implemented in the business strategy of a few hotels (in particular those that do business within international hotel chains) while tourist agencies do not show the change of their orientation towards socially more responsible behaviour (except for the Dubrovnik based Gulliver travel agency that is a part of the world's leading travel company, TUI Travel Plc.)

Finally, within the discourse of social sustainability, discussion on new trends of business performance in the partnership between public and private sector seems to be inevitable.

Public–private partnership (PPP) describes a government service or private business venture which is funded and operated through a partnership of government and one or more private sector companies. Public–private partnership involves a contract between a public sector authority and a private party, in which the private party provides a public service or project and assumes substantial financial, technical and operational risk in the project. A private sector consortium forms a special company called to develop, build, maintain and operate the asset for the contracted period. The increase of the public–private partnership projects has been the result of the processes related to the change of the government's role in the process of development (new forms of governance). As far as tourism industry is concerned, this concept has already been widely used across the world due to the fact that tourism business uses a great deal of public goods and government's role is to protect them. Following are the areas of tourism business where public–private partnership most often occurs:

- The tourist destination attractiveness' enhancement (infrastructure, new attractions and accommodation facilities, etc.);
- Marketing efficiency enhancement (development of the new product, promotion, new information/distributive systems);

- Productivity rise-up (through education, quality management, implementation of the new management techniques and new technologies);
- Enhancement of the management models (through education of all tourism stakeholders, implementation of the new management tools and concepts etc.)

As far as Croatian experience regarding private-public partnership in tourism is concerned there are still very few such examples. One of them, Suncani Hvar – ORCO hotel company as the first partnership project between the national government, local municipality authorities and an international hotel company has failed, and the company is struggling to survive overburdened by many unsolved problems. The project of health tourism resort in the Krapina County is to be realized through public-private partnership, as well as Posedarje Rivijera, a greenfield project, aimed at development of a high quality tourist resort. In 2009, Ministry of Tourism entered into partnership with 12 hotels and four chambers of commerce with the aim to subsidize scholarship for 320 pupils and 20 students. After finishing school they will work for the hotels that have entered the partnership. Furthermore, the Split-Dalmatian County has entered into the partnership with the owners of the real estates in the abandoned or devastated villages in the Dalmatian hinterland with the purpose of creating so called „eco-ethno villages". The County authorities are obliged to make infrastructural adjustments, development studies, management plans, etc. On the other side the real estate owners have to organize themselves into a non-governmental organisation which will represent their interests in the process of negotiating the terms under which they will put their real estates into function.

As could be seen, public-private partnership projects in Croatian tourism have gradually started to get introduced just recently and not much evidence on their presence has been recorded so far.

To conclude: Though positive changes have been recorded related to empowerment of individuals and communities to manage their own future, there is still much to do in terms of building social capital at all the three levels, i.e. macro, mezzo and micro level. When improvements in this area happen, changes in other areas of sustainability are expected to get realized more easily.

5. Conclusion

The nature of tourism is obviously ambivalent. On the one hand, it might be a valuable source of income and employment, potentially acting as a catalyst for wider socio-economic development or regeneration. On the other hand, the growth and expansion of tourism generates different environmental costs related to different types of degradation, misallocation or destruction of natural resources. These are usually accompanied by a variety of economic, social, cultural and political consequences. Hence it is obvious that in the absence of appropriate management techniques and tools, tourism has the ability to destroy the very resources upon which it depends. Without strategic approach to its development and the use of integral planning to this matter, fulfilment of sustainability principles is threatened. By researching the case of Croatian tourism model of development we have shown that despite being recognized institutionally, sustainability has not been achieved in any of the areas under study. By this the main hypothesis of the chapter has been proved. Being qualitative by its nature this research is partially leaning on the author's subjective opinion. However, the author has tried to the best of her abilities to consult as many relevant sources of literature and information as possible. As compared to other similar case studies most of which are focused on a particular area or an aspect of sustainability, this one has tried to cover all the three of them thus getting a holistic dimension. Namely, the three aspects/areas of sustainability (i. e., economic, ecological and social) are all interlaced to such an extent that it is hard to say where one ends and the other one begins. However the social area sustainability seems to be a starting point for better understanding of the reasons for possible failures in achieving sustainability in two other areas (i.e. economic and environmental ones).

Though this research as any other one could have been done in a different manner, and including more relevant issues, we believe that even as such it has 'unmasked' the particular tourism development model that is very often named sustainable or at least 'nature friendly'. It would certainly be of more help if some additional, more concrete indicators could have been presented. However, since they are usually done for the level of a community or a region (not a country) it was not possible to do so. There have been mostly general trends that were analyzed.

Finally, in order to give this research a bit of pragmatism, few practical proposals are to be given, aiming at putting sustainability principles in life. First of all, we believe that the popular dilemma of whether mass tourism in Croatia is needed or not is quite out of place. It is clear that tourism will not lose its mass character, indeed quite the opposite. Therefore the real dilemma lies not in whether we need mass tourism or not, because it will remain a mass phenomenon according to all the indicators, but rather what kind of mass tourism do we actually need and want? If Croatia continues to focus only on boosting the number of tourists, then it is positioned for the continuation of the current development trend of mass and undifferentiated tourism. On the other hand, we could opt for modest growth in the number of tourists but focus on their seasonal and regional redistribution, as well as increased profitability within a sustainable environmental model. To this end Croatia has to create a spectrum of tourist services that can satisfy different kinds of guests, distinguished not just by their purchasing power, but by their different affinities and habits. It will also mean a trickling down of tourism industry benefits to many more businesses and Croatian residents. With this second option in mind, the following strategic measures ought to be adopted:

-    The level of economic development in all Croatian regions should be increased, thus creating preconditions for development of economically more sustainable tourism. It is a quite common knowledge that tourism cannot be developed unless national economy is appropriately developed. Consequently, it has to be understood and accepted by the public that is not sole tourism that could solve the problem of underdevelopment, which is the idea quite often placed by Croatian authorities;
-    Development of modern technologies (especially information and communication ones) should be promoted together with the new business methods based on knowledge;
-    Domestic producers should be included in all the direct and indirect segments of tourism supply chain (agricultural production, fishery, construction etc.) in order to reduce the dependence of the overall tourism industry on imports. By this development

of the recognizable "Croatian brand" would be enhanced and well-being of Croatian citizens fostered;

- Incentives aimed at reduction of the grey economy in the small tourism business sector should be promoted;
- Awareness of the positive and negative effects of tourism should be raised among local population;
- Local communities should be empowered to affect and manage their future. To this end different models/methods of empowerment enhancement could be used. Hence, leaning on the examples of good practices from the UK, Timothy (2002) mentions Gill's idea published in 1996, known as 'living room meetings', which involves informal gatherings of small groups of community members in a moderated, yet relaxed situation throughout the community. He also explains benefits of the Fitton's ' planning for real' method which is a form of town meeting that involves bringing the community together before the planning process begins. Another method that has found considerable success is through household questionnaires, whose benefits were already explained in 1994 by Simmons. These methods help identifying issues that are important to an area, focus on the needs of the community and highlight opportunities for improvement. It gives everyone in the community an opportunity to participate and encourages them to think about tourism, local issues and the environment in depth (Timothy, 2002), or help spreading, as Porter said "social glue" (Porter, 1998).
- Local authorities should support promotion and implementation of a planning solution which ensures that the unique identity of the destination is maintained;
- The level of control over the behaviour of all the relevant stakeholders should be raised;
- Management efficiency of the tourism system in general has to be enhanced through education, partnership, networks and different other types of cooperation;
- Principles of responsible ecological behaviour on all levels should be promoted and implemented more intensely by introducing eco certificates, codes of conduct, Carrying Capacity Assessment and other tools and instruments aimed to this matter;
- Parallel with the increase in diversity and quality of tourist supply various tourist segments ought to be attracted and distributed more evenly throughout the country with the help of different economic and management instruments, such as price differentiation, marketing and demarketing techniques etc. Thus, saturated coastal areas could be relieved in favour of rural and inland destinations hence making the overall tourism more sustainable.

This list is not exhaustive and indicates a range of principles that underpin strategic and integrated planning for tourism areas. As the matter of sustainability is a very broad area of research, any of the elaborated principles could be an object of some future research.

To conclude: it is important to stress, yet again, that there are no institutional or practice models from elsewhere which can be transplanted in Croatia as a kind of panacea promoting sustainable tourism development. Rather, as Stubbs has pointed (as cited in Petrić, 2007), what is needed is the creation of networks, arenas and spaces, locally, nationally, and internationally, for exchanges of experiences and the elaboration of good practice, not in terms of set formulae, but in terms of attempting to grapple with the reason why certain initiatives appear to have had positive effects and others less so.

6. References

Analiza turističke godine 2010, (April 2011). 26.06.2011, Available from:
    http://www.mint.hr/UserDocsImages/110530-analiza-2010-w.pdf,   Ministarstvo
    turizma, Zagreb

Blažević, B. (2007). *Turizam u gospodarskom sustavu*, Fakultet za turistički i hotelski
    menadžment, ISBN 978-953-6198-96-2, Opatija

Bramwell, B. & Lane, B. (1993). Sustainable Tourism; An Evolving Global Approach. *Journal
    of Sustainable Tourism*, Vol. 1, No.1, (January, 1993), pp. 1-5, ISSN 0966-9582

Bramwell, B. & Lane, B. (Eds). (2000). *Tourism Collaboration and Partnerships, Politics, Practice
    and Sustainability*, Channel View Publications, ISBN 1-873150-79-2, Clevedon

Butler, R. (1999). Sustainable Tourism: A State-of-the-Art Review. *Tourism Geographies*, Vol.
    1, No. 1., (February 1999), pp. 7-25, ISSN: 1470-1340, Retrieved from
    http://dx.doi.org/10.1080/14616689908721291

Census 2001., 5.06.2011. Available from: http://www.dzs.hr/default_e.htm, Croatian
    Bureau of Statistics

Climate Change and Tourism – Responding to Global Challenges. (2008). 18.05.2011,
    Available from:
    http://www.unwto.org/sdt/news/en/pdf/climate2008.pdf,   World   Tourism
    Organization and United Nations Environment Programme

Coccossis, H., Mexa, A. & Collovini, A. (2002). *Material for a Document "Defining, Measuring
    and Evaluating Carrying Capacities in European Tourism Destination"*. University of
    Aegean, Department of Environmental Studies, Athens. 15.05.2011, Available from:
    http://ec.europa.eu/environment/iczm/pdf/tcca_material.pdf

Croatia-Travel and Tourism Climbing to the new Highlights, Travel & Tourism Economic
    Research. (2006). World Travel and Tourism Council, London

Demunter, C. (2010). Statistics in focus, *Eurostat*, No. 54 (November 2010), 16.05.2011,
    Available from:
    http://epp.eurostat.ec.europa.eu/cache/ITY_OFFPUB/KS-SF-10-054/EN/KS-SF-
    10-054-EN.PDF

Dragičević, M., Klarić Z. & Kušen, E. (1997). *Guidelines for Carrying Capacity Assessment for
    Tourism in Mediterranean Coastal Zones*, Priority Actions Programme, Regional
    Activity Centre, Split, Retrieved from:
    http://www.pap-thecoastcentre.org/pdfs/Guidelines%20CCA%20Tourism.pdf

Dunn, B. (1995). Success Themes in Scottish Family Enterprises; Philosophies and Practices
    Through the Generations, *Family Business Review*, Vol. 8, No 1, (March 1995), pp. 17-
    28, ISSN: 0894-4865

Franičević, V. & Bartlett W. (2001). Small Firms Networking and Economics in Transition; an
    Overview of Theories, Issues and Policies. *Zagreb International Review of Economics
    and Business*, Vol. 4, No.1, (November, 2001), pp. 63-89., ISSN 1331-5609

Fredotović, M., Maroević M., Madiraca, M., Nikolić, T., Alegro, A., Kralj, J., Bakran-
    Petricioli, T. & Biljaković, M. (2003). *Conservation and Sustainable Use of Biodiversity
    in the Dalmatian Coast through Greening Coastal Development, Concept paper for a full
    sized GEF project*. Ministry of Environmental Protection, Physical Planning and
    Construction and the United Nations Development Programme/Global
    Environment Facility, Zagreb

Godišnje izvješće 2009 – Gospodarski odnosi s inozemstvom, (2010). 14.05.2011, Available from: http://www.hnb.hr/publikac/godisnje/2009/h-god-2009.pdf,Hrvatska narodna banka

Hall, C. M. & Lew A. A. (1998). The Geography of Sustainable Development; an Introduction, In: *Sustainable Tourism: A Geographical Perspective*. Hall C. M. & Lew A. A. (eds.), pp. 1-12, Pearson Education Limited, ISBN-10: 0582322626, Harlow

Hall, D. (2000). Sustainable Tourism Development and Transformation in Central and Eastern Europe. *Journal of Sustainable Tourism*, Vol. 8, No. 6, (n. d.), pp. 441-457, ISSN: 0966-9582

Haywood, K. M. (1988). Responsible and Responsive Tourism Planning in the Community. *Tourism Management*, Vol. 9, No. 2, (June 1988), pp. 105-118, ISSN: 02615177

Horak, S., Marušić, Z., Krasić, D., Tomljenović, R., Kušen, E.,Telišman-Košuta, N. & Carić, H. (2007). *Studija održivog razvoja cruisinga u Hrvatskoj*, Institut za turizam, Retrieved from: http://www.mint.hr/UserDocsImages/SAZETAK-Studija-kruzing.pdf

Hrvatska u brojkama, (2010). 25.06.2011, Available from: http://www.dzs.hr/Hrv_Eng/CroInFig/hrvatska_u_brojkama.pdf

Implementation of Agenda 21: Review of progress made since the United Nations Conference on Environment and Development in 1992, Republic of Croatia-Country Profile, (April 1997). 25.05.2011, Available from: http://www.un.org/esa/earthsummit/croat-cp.htm, Government of the Republic of Croatia, Zagreb

Inskeep, E. (1991). *Tourism Planning: an Integrated and Sustainable Development Approach*, Van Nostrand Reinhold, ISBN-10: 0442001223, New York

Inskeep, E. (2006). *National and Regional Tourism Planning, Methodologies and Case Studies*, World Tourism Organisation, ISBN 10: 92-844-0727-3, Madrid

Ivandić, N., Telišman-Košuta, N. & Ledić-Blažević, M., (2010). *Hrvatsko hotelijerstvo 2008, Poslovanje hotelskih poduzeća*, 10.06.2011, Available from: http://hgk.biznet.hr/hgk/fileovi/17427.pdf, Institut za turizam

Jurčić, Lj. (2000). The import dependence of Croatian tourism, *Acta turistica*, Vol. 12, No. 1, (June 2000), pp. 1-16, ISSN: 353-4316

Kljenak, N. (2011). *Ljeto 2010 – ljeto preokreta za iznajmljivače*, 26.06.2011, Available from: http://www.turizaminfo.hr/ljeto-2010-ljeto-preokreta-za-iznajmljivace-na-crno

Law on Agricultural Land (Zakon o poljoprivrednom zemljištu). (2008). *Narodne novine No. 125/08*; Retrieved from: http://narodne-novine.nn.hr/clanci/sluzbeni/2008_12_152_4145.html

Law on Environment Protection (Zakon o zaštiti prirode), *Narodne novine, No. 82/1994, 110/2007*, Retrieved from: http://narodne-novine.nn.hr/clanci/sluzbeni/288893.html

Law on Golf Courses (Zakon o igralištima za golf). (2008). *Narodne novine, No. 152/08*, Retreived from: http://narodne-novine.nn.hr/clanci/sluzbeni/2008_12_152_4146.

Mansfeld, Y., (2002). Ponovno otkrivanje destinacije pomoću "mrežnog" sustava destinacijskog menadžmenta: primjer sjevernog, ruralnog dijela Izraela, *Turizam* Vol. 4, No. 50, (n. d.), pp. 361-371, ISSN: 1332-7461

Marušić, Z., Čorak, S., Hendija, Z. & Ivandić, N., (2008). *TOMAS ljeto 2008, Stavovi i potrošnja turista u Hrvatskoj*, Institut za turizam, ISBN: 978-953-6145-19-5, Zagreb

Marušić, Z., Čorak, S., Sever, I. & Ivandić, N. (2010). *TOMAS ljeto 2010, Stavovi i potrošnja turista u Hrvatskoj,* Institut za turizam, ISBN: 978-953-6145-23-2, Zagreb

Mason, P., Johnston, M. & Twynam, D. (2000). The World Wide Fund for Nature Arctic Project. In: *Tourism Collaboration and Partnerships, Politics, Practice and Sustainability,* Bramwell B., Lane B. (eds.), pp. 98-116, Channel View Publications, ISBN 1-873150-79-2, Clevedon

Mason, P. (2003). *Tourism Impacts, Planning and Management.* Elsevier, Butterworth and Heinemann, ISBN: 0 7506 5970X, Oxford

Mathieson, A. & Wall, G. (1982). *Tourism: Economic, Physical and Social Impacts.* Longman Group, ISBN: 0582300614, Harlow

Munasinghe, M. (2003). *Analysing the Nexus of Sustainable Development and Climate Change: an Overview,* Organisation for Economic Cooperation and Development (OECD), Retrieved from, http://www.oecd.org/dataoecd/32/54/2510070.pdf

Murphy, P. E. (1985). *Tourism, a Community Approach.* Methuen, ISBN: 0416 35930 2, New York

Nordin, S., (2003). *Tourism Clustering and Innovation, Path to Economic Growth and Development,* European Tourism Research Institute, Östersund, Retrieved from: http://ekstranett.innovasjonnorge.no/Arena_fs/tourism-clustering-and-inno_etour0104.pdf

Novak, M., Petrić, L. & Pranić, Lj., (2011). The effects of selected macroeconomic variables on the presence of foreign hotels in Croatia, *Tourism and Hospitality Management,* Vol. 17, No. 1, (June 2011), pp. 45-6), ISSN: 1330-7533

Perica, S. (2011). *U Hrvatskoj 150.000 bespravnih objekata,* 26.06.2011, Available from: http://www.vecernji.hr/vijesti/u-hrvatskoj-150-000-bespravnih-objekata-clanak-295867

Petrić, L. (2003). Socioekonomska analiza Dubrovačko-neretvanske i Splitsko-dalmatinske županije s posebnim naglaskom na analizu razvojnih uvjeta i mogućnosti turizma na otocima Visu, Mljetu i Lastovu u kontekstu očuvanja bioraznovrsnosti. In the project: *Ecoregional Action Programme, Middle and South Dalmatian Islands and Coast.* World Wildlife Fund/Mediterranean Programme, Zelena akcija, Zagreb & Sunce, Split

Petrić, L &, Mrnjavac, Ž. (2003). Tourist Destination as a Locally Embedded System - Analogy Between Theoretical Models of Tourist Destination and Industrial District. *Turizam,*Vol. 51, No. 4, (n.d.), pp. 375-383, ISSN: 1332-7461

Petrić, L., Fredotović, M., Grubišić, D. & Baučić, M., (2004). *Program održivog razvitka otoka - otočne skupine Prvić i Zlarin,* Ekonomski fakultet Split & Ministry of Sea, Transport, Tourism and Development, Zagreb

Petrić, L. (2005). Sectoral study on tourism development at regional/local level. In the project: *Conservation and Sustainable Use of Biodiversity in the Dalmatian Coast through Greening Coastal Development - COAST,* Ministry of Environmental Protection, Physical Planning and Construction, United Nations Development Programme & Global Environment Facility, Zagreb

Petrić, L. (2006). Tourism in Croatia: The Spanish or Costa Rican Road?, *Development and Transition,* No. 1, (July 2006), pp. 20-26, United Nations Development Programme, Zagreb

Petrić, L. (2006). Challenges of Rural Tourism Development: European Experiences and Implications for Croatia, *Acta Turistica*, Vol. 18, No. 2, (December, 2006), pp.138 – 170, ISSN 0353-4316

Petrić, L. (2007). Empowerment of Communities for Sustainable Tourism Development, Case of Croatia, *Turizam*, Vol. 55, No. 4, (n.d.), pp. 431-443, ISSN:1332-7461

Petrić, L. (2008). How to Develop Tourism Sustainably in the Coastal Protected Areas? The Case of Biokovo Park of Nature, Croatia, *Acta Turistica Nova*, Vol. 2, No. 1, (June 2008), pp. 5-24, ISSN:1846-4394

Petrić, L. & Pranić, LJ. (2009). Croatian Hoteliers' Attitudes Towards Environmental Management, *Proccedings of the 8th International Conference Challenges of Europe: Financial Crisis and Climate Change, PDF on CD ROM with full papers*, ISSN 1847-4497, Split- Bol (island of Brač), May 2009

Petrić, L. & Pranić, Lj. (2010). The Attitudes of the Island Local Community Towards Sustainable Tourism Development – the Case of Stari Grad, Island Hvar, *Proceedings of the 1st International Conference on Island Sustainability*, ISSN 1743-3541, Bol/Brač, September, 2010

Pivčević, S. & Petrić, L. (2011). Empirical Evidence on Innovation Activity in Tourism - The Hotel Sector Perspective, *The Business Review*, Cambridge, Vol. 17, No. 1, (May 2011), pp. 142-148, ISSN 1553 - 5827

Porter, M. E. (1990). *The Competitive Advantage of Nations*, The Free Press, ISBN: 0-684-84147-9, New York

Porter, M. E. (1998). Clusters and the New Economics of Competition, *Harvard Business Review*. (November-December 1998), pp. 77-90, 2.07.2011, Available from: http://www.wellbeingcluster.at/magazin/00/artikel/28775/doc/d/porterstudie.pdf?ok=j

Poslovanje hotela u Hrvatskoj 2009, (2010). 16.06.2011, Available from: http://www2.hgk.hr/komora/hrv/zupkom/split/hgkcms/images/Analiza_hote lijerstvo_finalno.pdf, Horwath HTL

Report of the World Commission on Environment and Development: Our Common Future, (1987). 4.05.2011, Available from http://www.un-documents.net/wced-ocf.htm , World Commission on Environment and Development

Sharpley, R. (2009). *Tourism Development and the Environment: Beyond Sustainability?* Earthscan, ISBN: 978-1-84407-733-5, London

Simmons, D. (1994). Community participation in tourism planning. *Tourism Management*, Vol. 15, No. 2, (April, 1994), pp. 98-108, ISSN: 02615177

Škrabalo, M., Miošić-Lisjak, N. & Bagić, A. (2007). Baseline study on Corporate Social Responsibility in Croatia, for the project: *Accelerating CSR practices in the new EU member states and candidate countries as a vehicle of harmonization, competitiveness and social cohesion"*, United Nations Development Programme, ISBN 978-953-7429-10-2, Zagreb, 29.06.2011, Available form: http://www.undp.hr/upload/file/205/102972/FILENAME/DOP_english.pdf,

Sofield, T. H. B. (2003). *Empowerment for Sustainable Development*, Pergamon Press, ISBN: 0-08-043046-2, Oxford.

Southgate, C. & Sharpley, R. (2002). Tourism Development and the Environment: Beyond Sustainability; In: *Tourism and Development, Concepts and Issues*, Sharpley, R., Telfer & D. J. (eds), (pp. 231-262), Channel View Publications, ISBN: 1-873150-34-2, Clevedon

Strategija razvoja nautičkog turizma RH za razdoblje 2009-2019. (2008). 10.06.2011, Available from: http://www.mint.hr/UserDocsImages/081224-61_01.pdf,   Ministarstvo mora, prometa i infrastrukture & Ministarstvo turizma, Zagreb

Stubbs, P. (2007). Aspects of Community Development in Contemporary Croatia: Globalisation, Neo-liberalisation and NGO-isation. In: *Revitalising Communities,* Dominelli, L. (ed), pp. 161-174, Ashgate Publishing, ISBN: 0754644987, Aldershot

Šutalo, K. (2009). *Kupac zemlje na Srđu odustaje od golf terena,* 26.06.2011, Available from: http://www.monitor.hr/clanci/kupac-zemlje-na-srdu-odustaje-od-golf-terena/22468/

Taylor, T., Fredotović, M., Povh, D. & Markandya, A. (2005). Sustainable Tourism and Economic Instruments: International Experience and the Case of Hvar, Croatia, In: *The Economics of Tourism and Sustainable Development,* Lanza, A., Markandya, A. & Pigliaru, F. (eds.), pp. 197-224, Edward Elgar Publishing Limited, ISBN: 1845424018, Cheltenham, UK - Northampton, USA

Telfer, D. J. (2002). The Evolution of Tourism and Development Theory, In: *Tourism and Development, Concepts and Issues,* Sharpley, R. & Telfer, D. J. (eds), pp. 35-79 Channel View Publications, ISBN: 1-873150-34-2, Clevedon

Timothy, D. J. (1998). Cooperative Tourism Planning in a Developing Destination. *Journal of Sustainable Tourism,* Vol. 6, No. 1, (n. d.), pp. 52-68, ISSN 0966-9582

Timothy, D. J. (2002). Tourism and Community Development Issues. In: *Tourism and Development, Concepts and Issues,* Sharpley, R. & Telfer, D. J. (eds), pp. 149-162, Channel View Publications, ISBN: 1-873150-34-2, Clevedon

Tomljenović, R., Marušić, Z., Weber, S., Hendija, Z. & Boranić S. (2003). *Strategija razvoja kulturnog turizma: Od turizma i kulture do kulturnog turizma.* Institut za turizam, Zagreb, Retrieved from: http://web.efzg.hr/dok/TUR//Strategija%20Razvoja%20Kulturnog%20Turizma. pdf

Tosun, C. (2000). Limits to Community Participation in the Tourism Development Process in Developing Countries. *Tourism Management,* Volume 21, Issue 6, (December 2000), pp. 613-633, ISSN: 02615177

Tourism Highlights, (2011). 11.05.2011, Available from: http://mkt.unwto.org/en/content/tourism-highlights,   United   Nations   World Tourism Organisation, Madrid

Tourism Satellite Account for Croatia (2011). 15.05.2011, Available from: http://www.wttc.org/eng/Tourism_Research/Economic_Data_Search_Tool/ World Travel and Tourism Council

Vehovec, M. (2002). Evolucijsko-institucionalan pristup razvoju poduzetništva, In: *Poduzetništvo, institucije i sociokulturni kapital,* Čengić, D. & Vehovec, M. (eds.), pp. 13-36, Institut društvenih znanosti "Ivo Pilar", ISBN: 953-6666-25-1, Zagreb.

Vukonić, B. (2005). Study on tourism development at national level, In the project: *Conservation and Sustainable Use of Biodiversity in the Dalmatian Coast through Greening Coastal Development - COAST,* Ministry of Environmental Protection, Physical Planning and Construction, the United Nations Development Programme & Global Environment Facility

Weaver, D. (2006). *Sustainable Tourism,* Elsevier Butterworth- Heinemann, ISBN: 0 7506 6438 X, Oxford

8

An Approach to Sustainable Development by Applying Control Science

Kazutoshi Fujihira
Institute of Environmentology
Japan

1. Introduction

Nowadays, humankind is facing various environmental and social problems; for example, global warming, the destruction of ecosystems, an increase of areas where water supplies are insufficient, the tight supply-demand situation for oil and metals, poverty, economic crises and conflicts. It is our ultimate goal as humans to solve or prevent environmental and social problems and achieve "sustainable development" or "sustainability."

In order to solve or prevent such problems and achieve sustainable development, "human beings must **control** their activities appropriately," I wrote in my books titled *An Introduction to Environmentology* (Fujihira, 1999) and *A Short Introduction to Environmentology* (Fujihira, 2001). In 2001, when I published the second book, I conceived the idea of applying the science of "control" to this ultimate challenge of humankind. "Control" is generally defined as "purposive influence toward a predetermined goal" (Beniger, 1986). Moreover, control science can be applied to all goal-oriented tasks. In fact, control science is applied to a variety of fields such as engineering, economics, agriculture, and medicine; especially control engineering has a long history and has produced remarkable results. Accordingly, it is a rational and reliable approach to apply control science to the task of achieving sustainable development.

Quickly realizing this point, I started conducting research. After that, I obtained cooperation from experts, including a leading scientist in control engineering. The finished research has shown the basic control system for sustainable development and an educational methodology for sustainable development with case studies (Fujihira et al., 2008; Fujihira & Osuka, 2009). The results of the case studies have demonstrated the validity of that basic control system as well as that educational methodology.

Recently we have aimed to show a methodology of designing practical control systems for sustainable development. Here this study, as the first step, discusses a method for promoting smooth design of such control systems. Chapter 2 again shows the basic control system for sustainable development. Chapter 3 provides the two-step preparatory work for smooth control system design. In Chapter 4, we apply this method to homes and demonstrates a case study. Chapter 5 examines the results of this case study and shows the effectiveness of this method.

2. The basic control system for sustainable development

Fig. 1 shows the basic control system for sustainable development (Fujihira et al., 2008; Fujihira & Osuka, 2009). 'Controlled objects' are human activities which cause environmental or social problems; the units of human activities are various. 'Controlled variables' are the variables that relate to the human activities and need to be controlled for solving or preventing the problems. 'Disturbances' are harmful influences on controlled objects which are caused by environmental and social problems. Examples of the disturbances are damage caused by environmental pollution, flood or landslide damage resulting from unbridled land development, and various kinds of damage caused by global warming.

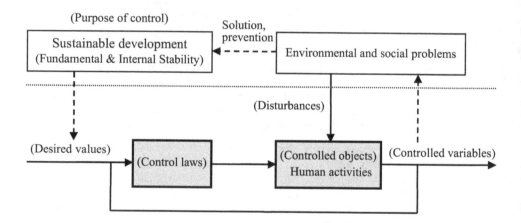

Fig. 1. The basic control system for sustainable development

Desired values are derived from the purpose of control, that is to say, sustainable development. The model of sustainable development (Fig. 2) demonstrates that sustainable development requires both 'Fundamental Stability' and 'Internal Stability,' in order to accomplish the long-term well-being of all humankind, or ultimate end, within the finite global environment and natural resources, or absolute limitations (Fujihira et al., 2008; Fujihira & Osuka, 2009). 'Fundamental Stability' means environmental stability and a stable supply of natural resources; the conditions for Fundamental Stability are environmental preservation and the sustainable use of natural resources. On the other hand, 'Internal Stability' means social and economic stability; the conditions for Internal Stability are health, safety, mutual help and self-realization, which are essential for well-being of humans. In addition, natural science, social science and human science, which are placed between Absolute Limitations, Fundamental Stability, Internal Stability and Ultimate End, are necessary to investigate the respective relationships.

The control objective is to adjust the controlled variables to their desired values. Furthermore, the control system requires designing and implementing 'control laws,' or measures for achieving the control objective.

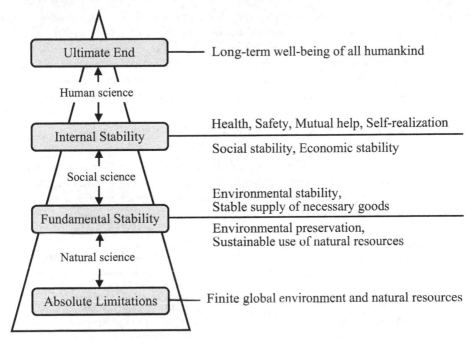

Fig. 2. The model of sustainable development

3. Two-step preparatory work for smooth control system design

There is a standard procedure that can be applied to the design of most control systems. Fig. 3 shows main steps in this procedure: 1) identifying a controlled object and control objective, 2) understanding the controlled object and control objective, 3) designing control laws, 4) implementing control laws. The first step "identifying a controlled object and control

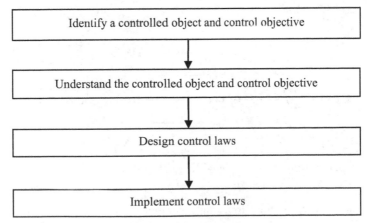

Fig. 3. Main steps in designing control systems

objective" requires designers of practical control systems for sustainable development to identify controlled variables and their desired values as well as a controlled object. Therefore, preparatory work for designing such control systems is primarily intended to identify these system components. This preparatory work consists of two steps: (1) determining the relationship between the standard human activities and sustainable development, (2) sustainability checkup on human activities as an object (Fujihira & Osuka, 2010, 2011).

3.1 Determining the relationship between the standard human activities and sustainable development

The first step aims to comprehensively determine the relationship between the standard human activities and sustainable development. The standard human activites means typical human activities among human activities which belong to one group and the same unit. Fig. 4 demonstrates the concept of this step.

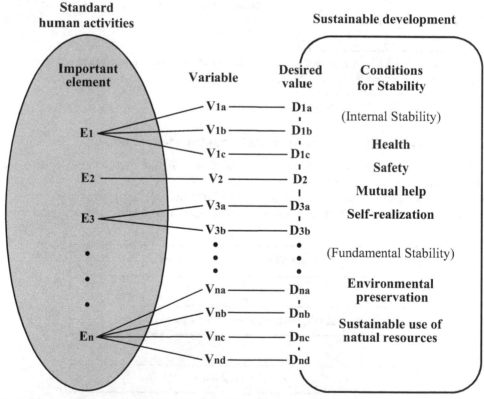

Fig. 4. The concept diagram of determining the relationship between the sandard human activities and sustainable development

The first step starts with selecting important elements from the standard human sctivites. Human activities in one group and the same unit include almost the same elements.

Accordingly, system designers first select such common elements from the standard human activities. In this connection, if system designers find one or more factors which influence the selection, the selection work will be more efficient. In addition, the elements whcih are expected to closely relate to sustainable development should always be added to a set of important elements, regardless of whether such elements are common or not. For example, when the 'building' is chosen as a unit of human activities, "equipment for harnessing natural energy" should be selected as an important element, even if it is uncommon in present ordinary buildings.

After selecting important elements, system designers determine the relationship between such elements and sustainable development. This work consists of three processes: 1) considering the relationship between each element and the conditions for both Fundamental Stability and Internal Stability, such as health, safety and environmental preservation; 2) identifying variables which can indicate the degree of stability; 3) setting the desired values of the variables that can achieve stability. As shown in Fig. 4, the number of variables which connect to one element is not necessarily one, but can be many. In addition, both identifying variables and setting desired values need to be conducted on the basis of the latest scientific knowledge, technology and social conditions.

3.2 Sustainability checkup on human activities as an object

In the second step, system designers conduct a sustainability checkup on human activities as an object. To be concrete, they first measure or estimate the aforementioned variables of human activities as an object. Next, they compare the measured or estimated values with the desired values and assess the degree of stability.

After the assessment, the variables that are lower than the desired values need to be identified as "controlled variables." The variables that fall substantially below the desired values are especially required to be identified as "controlled variables." In addition, human activities as an object which include one or more controlled variables are naturally identified as a "controlled object."

In addition, this sustainability checkup is applied to both "new" and already "existing" human activities. *Oxford Dictionary of English* defines 'activity' as "a thing that a person or group does or has done" (Oxford Dictionaries, 2010). In this context, "new" human activities are equivalent to "things that people or groups do," and "existing" human activities correspond to "things that people or groups have done." When the object of a sustainability checkup is new human activites, system designers conduct it by examining the plan or blueprint for such activities. On the other hand, when the object is existing human activities, system designers conduct a checkup by inspecting the actual human activities. In the latter case, if system designers can obtain the plan or blueprint of such activities, it is desireble to examine it as well as the actual human activities. Furthermore, if both new and existing human activities are checked and controlled for sustainable development, the goal will be achieved more smoothly.

4. Case study

We have conducted a case study, selecting the home as a unit of human activities. In this case, the preparatory work consists of two steps: 1) determining the relationship between the standard home and sustainable development, 2) sustainability checkup on a home as an object.

4.1 Determining the relationship between the standard home and sustainable development

In the first step, system designers need to select important elements of the standard home and determine the relationship between such elements and sustainable development.

4.1.1 Two factors on selecting elements of the standard home

In order to determine important elements of the standard home, we have analyzed two main factors in making our selection. They are "material" and "space," as shown in Table 1 (Fujihira, 2011). The first factor "material" regards the home as an object which contains material elements such as framework, exterior, interior and piping. Moreover, when observing the details of such material elements, they can be broken down further into smaller material elements; for example, framework includes pillars and beams. On the other hand, the other factor "space" regards the home as an object which consists of spatial elements such as rooms and areas. If regarding the home as a mass of rooms, we can find more specific spatial elements; for instance, a living room, dining room, kitchen and bedroom.

In this case study, we have observed the standard home based on both of these factors. As a result, we have determined important elements, as shown in the central column of Table 2-1 and Table 2-2. "Material elements" are from 'framework' to 'fence;' "spatial elements" are from 'rooms used at daytime' to 'garden area,' which are demonstrated in the bottom of Table 2-2.

Factor	Material	Space
Examples of elements (details)	(a) Framework (pillar, beam, etc.) (b) Exterior (roof, outer wall, etc.) (c) Interior (floor, inner wall, ceiling, etc.) (d) Piping (water pipe, gas pipe, etc.)	(a) Room (living room, dining room, bedroom, kitchen, bathroom, etc.) (b) Area (garden area, exterior area, etc.)

Table 1. Two factors on selecting elements of the home

4.1.2 Relationship between the standard home and sustainable development

After selecting the elements of the standard home, we have determined the relationship between these elements and sustainable development. The left side of Table 2-1 and Table 2-2 shows the relationship between the elements and Fundamental Stability; the right side demonstrates the relationship between the elements and Internal Stability. Considering the relationship between each element and the stability conditions, we have identified variables which indicate the degree of stability. In addition, we have set the desired values of these variables that can achieve stability.

The rest of this section briefly describes the relationship between each element and sustainable development, in order from the top of Table 2-1.

[Material elements]

- ### Framework

Considering the relationship between "framework" and 'sustainable use of natural resources,' a condition for Fundamental Stability, we have identified 'durability' and 'raw materials' as variables. The desired values of 'durability' and 'raw materials' are the 'deterioration resistance grades' of the *Japan Housing Performance Indication Standards* (JHPIS) (Japanese Ministry of Land, Infrastructure and Transport, 2001), and the 'assessment levels of resources saving' of *CASBEE for Home*, or *Comprehensive Assessment System for Building Environmental Efficiency for Home (Detached House)* (Japan GreenBuild Council & Japan Sustainable Building Consortium, 2008), respectively.

On the other hand, considering the relationship with 'safety,' a condition of Internal Stability, we have selected 'resistance to earthquakes' and 'wind resistance' as variables, and the 'seismic resistance grades' and the 'wind-resistant grades' of the JHPIS as their desired values (Japanese Ministry of Land, Infrastructure and Transport, 2001). In Japan the strength of framework against earthquakes is regarded as extremely important since Japan is a major quake-prone country.

Furthermore, in areas of heavy snowfall, 'resistance to snowfall' needs to be included as a variable, although it is excluded from the table. In this way, variables and their desired values can be changed or varied with the surrounding environment.

- ### Exterior

As for "exterior," which includes roofs and outer walls, we have selected 'durability,' 'raw materials' and 'sunlight reflectivity' as variables relating to Fundamental Stability. 'Raw materials' is excluded from the table, for reasons of space. The desired value of sunlight reflectivity has been set at '0.3 or over' because good sunlight reflectivity prevents the exterior of homes from accumulating the heat of sun and leads to the mitigation of the heat island phenomenon. On the other hand, we have identified 'fire resistance,' 'shape' and 'color' as variables relating to Internal Stability. In order to restrain the spread of fire, the exterior needs to satisfy a high fire resistance grade. Meanwhile, the shape and color of the roof and walls are necessary to harmonize with the surrounding landscape.

- ### Thermal insulation

We have identified 'thermal insulation performance' and 'raw materials' as the variables of "thermal insulation." 'Raw materials,' which relates to Fundamental Stability, is excluded from the table, for reasons of space. 'Thermal insulation performance' is especially significant since it relates to both Fundamental Stability and Internal Stability. An increase in thermal insulation performance leads to environmental preservation and sustainable use of natural resources through a decrease in energy usage for air conditioning and heating. Meanwhile, it also promotes the health of occupants through the stabilization of the indoor temperature. The desired value of thermal insulation performance has been set at the highest grade in the "Energy-Saving Action Grades" of JHPIS (Japanese Ministry of Land, Infrastructure and Transport, 2001). In addition, there are six area classifications for the thermal insulation standards for specific values in Japan, depending on the climate.

Stability condition	Desired value	Variable	Element	Variable	Desired value	Stability condition
Sustainable use of resources	JHPIS Sec. 3-1: Grade 2 or over	Durability	Framework	Resistance to earthquakes	JHPIS Sec. 1-1: Grade 2 or over	Safety
Sustainable use of resources	CASBEE LR$_{H}$2 1.1: Level 4 or over	Raw materials	Framework	Wind resistance	JHPIS Sec. 1-4: Grade 1 or over	Safety
Sustainable use of resources	JHPIS Sec. 3-1: Grade 2 or over	Durability	Exterior (roof, outer wall, etc.)	Fire resistance	JHPIS Sec. 2-6: Grade 3 or over	Safety
Enviro-preserve	0.3 or over	Sunlight reflectivity	Exterior (roof, outer wall, etc.)	Shape / Color	Harmony with surrounding landscape	Health
Enviro-preserve / Sustainable use of resources	JHPIS Sec. 5-1: Grade 4	Thermal insulation performance	Thermal insulation	Thermal insulation performance	JHPIS Sec. 5-1: Grade 4	Health
Sustainable use of resources	JHPIS Sec. 3-1: Grade 1 or over	Durability	Windows & doors	Area of window openings	20% of the floor area or more	Health
Enviro-preserve	Consideration for natural lighting	Position & Shape	Windows & doors	Wind resistance	JIS Grade: S-2 or over	Safety
Sustainable use of resources	Consideration for ventilation	Position & Shape	Windows & doors	Measures to prevent intrusions	CASBEE Q$_{H}$1 2.6: Level 4 or over	Safety
Enviro-preserve / Sustainable use of resources	JHPIS Sec. 5-1: Grade 4	Thermal insulation performance	Windows & doors	Thermal insulation performance	JHPIS Sec. 5-1: Grade 4	Health
Enviro-preserve / Sustainable use of resources	JIS Grade: A-3 or over	Airtightness	Windows & doors	Sound insulation performance	JHPIS Sec. 8-4: Grade 2 or over	Health
Enviro-preserve / Sustainable use of resources	CASBEE Q$_{H}$1 1.1.2: Level 4 or over	Sunlight adjustment capability	Windows & doors	Sunlight adjustment capability	CASBEE Q$_{H}$1 1.1.2: Level 4 or over	Health
			Floor	Sound insulation performance	JHPIS Sec. 8-1&2: Grade 3 or over	Health
			Floor	Differences in level	No differences	Safety / Health

[Note] (1) JHPIS is an abbreviation for the Japan Housing Performance Indication Standards. (2) CASBEE means Comprehensive Assessment System for Building Environmental Efficiency for Home (Detached house) – Technical Manual 2007 Edition. (3) JIS is an abbreviation for Japanese Industrial Standard.

Table 2-1. Relationship between the standard home and sustainable development

Stability condition	Desired value	Variable	Element	Variable	Desired value	Stability condition
Sustainable use of resources	CASBEE LR$_H$2 1.4: Level 4 or over	Raw materials	Interior	Formalde-hyde emission	JHPIS Sec. 6-1: Grade 3	Health
Enviro-preserve	Insulated	Heat insulation	Bathtub			
Sustainable use of resources	JHPIS Sec. 4-1: Grade 3 or over	Consideration for maintenance	Piping			
Enviro-preserve	Header type	Type of piping	Hot-water piping			
Sustainable use of resources	Insulated	Heat insulation				
Enviro-preserve Sustainable use of resources	90% or more	Primary energy efficiency	Water heater			
Sustainable use of resources	CASBEE LR$_H$1 3.1: Level 4 or over	Water-saving functions	Water-using equipment			
Sustainable use of resources	10% or more of the total water usage	Rainwater usage	Equipment for rainwater use	Rainwater usage	10% or more of the total water usage	Health Safety (in crises)
Enviro-preserve Sustainable use of resources	100% or more	Energy-saving achievement rate	Lighting fixtures & appliances			
Enviro-preserve Sustainable use of resources	Energy usage of the whole home or more	Harnessed natural energy	Equipment for harnessing natural energy	Harnessed natural energy	Energy usage of the whole home or more	Health Safety (in crises)
Enviro-preserve	Indigenous species	Species	Garden plants	Fire resistance	High or mid fire resistance	Safety
Enviro-preserve Sustainable use of resources	Hedge or Resources-saving materials	Material	Fence	Form	Not blocking sight line	Safety
					Not blocking communication	Mutual help
Enviro-preserve Sustainable use of resources	Places receiving a lot of sunlight	Places in the home	Rooms used at daytime			
			Specified bedroom	Relation with toilet & bath	On the same floor	Health Safety
Enviro-preserve Sustainable use of resources	Building close together	Places in the home	Rooms where water is used			
			Doorways	Differences in level	No differences	Safety Health
				Width	80cm or more	
Enviro-preserve	40% or more	Ratio to the exterior area	Garden area			

Table 2-2. Relationship between the standard home and sustainable development

- **Windows and doors**

We have identified a large number of items as the variables of "windows and doors;" for example, an area of window openings, sunlight adjustment capability, thermal insulation performance, and sound insulation performance. It is necessary to obtain sufficient brightness and appropriate sunlight through windows. On the other hand, windows need sufficient thermal insulation performance and sound insulation performance. In this way, windows need to meet a variety of conflicting requirements, which indicates that designing windows is extremely difficult. Furthermore, in order to meet such a variety of requirements, related elements, such as glass, eaves, awnings, blinds, shutters, and curtains, are often required to work together.

- **Floor**

"Floors" require two variables, that is, 'sound insulation performance' and 'differences in level,' both of which relates to Internal Stability. The floor of the rooms is necessary to satisfy sufficient sound insulation performance against the noise from the upper floor. The other variable 'differences in level' need to be removed from the floor, in order to allow elderly and handicapped people to move around safely and lead a normal life. Recently this variable has become more important in Japan due to a rapidly aging society.

- **Interior**

"Interior," which includes a floor, wall and ceiling, requires 'formaldehyde emission' and 'raw materials' as its variables. Formaldehyde is a major harmful pollutant; therefore, the desired value of formaldehyde emission is set at the level which is harmless to the health of the occupants.

- **Bathtub**

We have attached importance to 'heat insulation' as a variable of the "bathtub" since insulated bathtubs can reduce heat loss of the hot water. This consideration results from a Japanese lifestyle, that is, taking a bath every day.

- **Piping**

"Piping," including drainage pipes, water pipes and gas pipes, need 'consideration for maintenance' as an important variable toward a long service life. The 'maintenance grades' of the JHPIS, which has been identified as the desired value, requires consideration for making maintenance easier, such as not burying piping under concrete and creating openings for cleaning and inspection (Japanese Ministry of Land, Infrastructure and Transport, 2001).

- **Hot-water Piping**

We have identified 'type of piping' and 'heat insulation' as the variables of "hot-water piping," both of which relates to Fundamental Stability. If 'header type' hot-water piping is used, normally the diameter of piping leading from the header to the faucets of sinks and baths can be reduced. As a result, wastage of hot water can be decreased, as compared with the front-end-branching type. Moreover, if hot-water piping is 'insulated,' heat loss is further reduced.

- **Water heater**

We have identified 'primary energy efficiency' as a key variable of the "water heater." The desired value of the primary energy efficiency has been set at '90% or more.' This level can be realized by utilizing high energy-efficient water heaters, including electric heat-pump water heaters.

- **Water-using equipment**

"Water-using equipment," including toilet bowls, faucets and shower heads, requires 'water-saving functions' as its key variable. The desired value, the water-saving assessment levels of CASBEE, can be satisfied if two or more water-saving efforts are adopted from the following four choices: water-saving type toilets, bathroom thermostat type water faucet plus water-saving shower head, dish washer, and other water-saving methods (Japan GreenBuild Council & Japan Sustainable Building Consortium, 2008).

- **Equipment for rainwater use**

If "equipment for rainwater use" is installed, it can reduce the quantity of water supply and contributes to sustainable use of natural resources. We have set the desired value of 'rainwater usage' at '10% or more of the total water usage.' Storing rainwater also contributes to health and safety in crises, by securing emergency water.

- **Lighting fixtures and home appliances**

Lighting fixtures and home appliances such as refrigerators and televisions need to be energy-saving devices. We have identified the variable of such appliances as the 'energy-saving standard achievement rate' and set their desired value at '100% or more.' Japan's energy-saving standard achievement rate for each appliance is open to the public in the manufacturers' catalogue and the latest "Energy Conservation Equipment Catalogue" of the Energy Conservation Center, Japan (Energy Conservation Center, Japan, n.d.).

- **Equipment for harnessing natural energy**

Concerning "equipment for harnessing natural energy" such as solar panels, we have identified 'harnessed natural energy' as a variable relating to Fundamental Stability, and 'energy usage of the whole home or more' as its desired value. This desired value means achieving self-sufficiency in energy. Equipment for harnessing natural energy also contributes to health and safety in crises, by generating emergency energy.

- **Garden plants**

We have determined 'species' and 'fire resistance' as the variables of "garden plants." If indigenous or local species of plants are selected, such selection contributes to preserving the region's ecological environment. On the other hand, highly fire-resistant plants are effective to prevent the spread of fire. In general, evergreen trees and plants with thick leaves which contain large amounts of water have high fire resistance.

- **Fence**

As for "fence," 'material' has been identified as a variable relating to Fundamental Stability, and ecological materials such as a hedge as its desired value. On the other hand, 'form' has been selected as a variable relating to Internal Stability and 'not blocking sight line' and 'not blocking communication' as its desired values. These selections are based on the following ideas: good visibility brings 'safety' through preventing crimes and face-to-face communication leads to 'mutual help' in local community.

[Spatial elements]

- **Rooms used at daytime**

"Rooms used at daytime," which usually include a living room and dining room, should be preferentially planned in places receiving a lot of sunlight in the home. Such arrangement is effective to reduce the energy for lighting by utilizing sunlight efficiently. On the other hand, rooms used only at night-time such as bedrooms can be planned in places with little sunlight.

- **Specified bedroom**

A "specified bedroom" means a bedroom which is used or expected to be used by elderly or wheelchair users. Such a room and the bathroom area should be arranged on the same floor. This arrangement enables such occupants to use the toilet and bath easily.

- **Rooms where water is used**

"Rooms where water is used" includes a kitchen, bathroom, toilet, and washing room. If these rooms are built close together, the total length of water piping and drainage piping can be reduced. Moreover, this consideration helps reduce heat loss from hot-water piping.

- **Doorways**

A "doorway" is a space where a door opens and closes. 'No differences in level' in doorways allow elderly and wheelchair users to pass through them smoothly. On the other hand, '80cm or more,' the desired value of the 'width' of a doorway, is suitable for movement of a wheelchair.

- **Garden area**

The "garden area" is an area with plants such as trees, shrubs, herbs, grasses, and vegetables. A larger garden area is favourable for environmental preservation, including mitigation of heat island phenomenon, and a higher level of biodiversity. We determined its variable as the 'ratio of the garden area to the exterior area,' and set its desired value at '40% or more.' In addition, the garden area includes any planted area not only on the ground but also on the roof.

4.2 Sustainability checkup on a home as an object

In the second step, system designers measure or estimate the variables of a home as an object and assess them by comparing with the desired values. Table 3-1 and Table 3-2 demonstrate an example of sustainability checkup on a home as an object; Fig. 5 shows the external appearance of the home on which the checkup was done. In this case, the checkup results have been assessed in three grades: A, B and C. "A" means that the variable reaches the desired value. "B" signifies that the variable falls below the desired value. "C" means that the variable falls substantially below the desired value.

Here I view the checkup results, choosing several elements from Table 3-1 and Table 3-2. As for "framework," two of the four variables, 'durability' and 'resistance to earthquakes' have been assessed at B because they are lower than the desired values. The 'performance' of the "thermal insulation" has been assessed at C since it falls substantially below the desired value. The "water heaters" used in this home are old-typed gas heaters and their 'primary energy efficiency' is much lower than the desired value; therefore, it has been assessed at C. The "natural energy" harnessed by "equipment for harnessing natural energy" has been assessed at C because such equipment is not installed. The two variables of "fence," 'material' and 'form,' have been assessed at A, for hedge and resources-saving materials are utilized and the form does not block both sight line and communication. The variable of the "specified bedroom" has been assessed at A because both the specified bedroom and bathroom area are placed on the same ground floor. The variable of the "rooms where water is used" has been assessed at A since the kitchen, toilet, bath and the place for a washing machine are close together.

In the above example, the object of sustainability checkup has been an existing home. When checkup is done on existing homes, it is desirable to examine both the actual home and its design drawings. On the other hand, this checkup method can be applied to homes which are planned or designed for the future. In the latter cases, planners and designers estimate the values of variables, by examining the scheme drawings or design drawings.

After the sustainability checkup, the variables that have been assessed at B or C need to be identified as "controlled variables." The variables assessed at C are especially required to be identified as "controlled variables." In addition, this home has naturally been identified as a "controlled object" because it includes controlled variables. Moreover, such a sustainability checkup table enables system designers to find at a glance the following: the elements which should be controlled, controlled variables and their desired values. Therefore, it enables them to understand both what should be controlled and the courses of control.

Fig. 5. The external appearance of the home on which a sustainability checkup was done

Relationship between the element and Fundamental Stability				Element	Relationship between the element and Internal Stability			
Desired value	Assess.	Measured or estimated value	Variable		Variable	Measured or estimated value	Assess.	Desired value
JHPIS Sec. 3-1: Grade 2 or over	B	40 years	Durability	Frame-work	Resistance to earthquakes	JHPIS Sec. 1-1: Grade 1	B	JHPIS Sec. 1-1: Grade 2 or over
CASBEE LR_{H2} 1.1: Level 4 or over	A	Domestic wood	Raw materials		Wind resistance	JHPIS Sec. 1-4: Grade 1	A	JHPIS Sec. 1-4: Grade 1 or over
JHPIS Sec. 3-1: Grade 2 or over	B	30 years	Durability	Exterior (roof, outer wall, etc.)	Fire resistance	JHPIS Sec. 2-6: Grade 3	A	JHPIS Sec. 2-6: Grade 3 or over
0.3 or over	A	0.55	Sunlight reflectivity		Shape / Color	Harmony with landscape	A	Harmony with landscape
JHPIS Sec. 5-1: Grade 4	C	JHPIS Sec. 5-1: Grade 1	Thermal insulation performance	Thermal insulation	Thermal insulation performance	JHPIS Sec. 5-1: Grade 1	C	JHPIS Sec. 5-1: Grade 4
JHPIS Sec. 3-1: Grade 1 or over	B	30 years	Durability	Windows & doors	Area of window openings	18% of the floor area	B	20% of the floor area or more
Consideration for natural lighting	A	Adequate consideration	Position & Shape		Wind resistance	JIS Grade: S-2	A	JIS Grade: S-2 or over
Consideration for ventilation	A	Adequate consideration			Measures to prevent intrusions	CASBEE Q_{H1} 2.6: Level 3	B	CASBEE Q_{H1} 2.6: Level 4 or over
JHPIS Sec. 5-1: Grade 4	C	JHPIS Sec. 5-1: Grade 2	Thermal insulation performance		Thermal insulation performance	JHPIS Sec. 5-1: Grade 2	C	JHPIS Sec. 5-1: Grade 4
JIS Grade: A-3 or over	A	JIS Grade: A-3	Airtightness		Sound insulation performance	JHPIS Sec. 8-4: Grade 1	B	JHPIS Sec. 8-4: Grade 2 or over
CASBEE Q_{H1} 1.1.2: Level 4 or over	B	CASBEE Q_{H1} 1.1.2: Level 3	Sunlight adjustment capability		Sunlight adjustment capability	CASBEE Q_{H1} 1.1.2: Level 3	B	CASBEE Q_{H1} 1.1.2: Level 4 or over
				Floor	Sound insulation performance	JHPIS Sec. 8-1&2: Grade 1	C	JHPIS Sec. 8-1&2: Grade 3 or over
					Differences in level	No differences	A	No differences

[Note] (1) JHPIS is an abbreviation for the Japan Housing Performance Indication Standards. (2) CASBEE means Comprehensive Assessment System for Building Environmental Efficiency for Home (Detached house) – Technical Manual 2007 Edition. (3) JIS is an abbreviation for Japanese Industrial Standard.

Table 3-1. An example of sustainability checkup on a home as an object

Relationship between the element and Fundamental Stability				Element	Relationship between the element and Internal Stability			
Desired value	Assess.	Measured or estimated value	Variable		Variable	Measured or estimated value	Assess.	Desired value
CASBEE LR$_H$2 1.4: Level 4 or over	A	CASBEE LR$_H$2 1.4: Level 4	Raw materials	Interior	Formaldehyde emission	JHPIS Sec. 6-1: Grade 3	A	JHPIS Sec. 6-1: Grade 3
Insulated	C	Not insulated	Heat insulation	Bathtub				
JHPIS Sec. 4-1: Grade 3 or over	B	JHPIS Sec. 4-1: Grade 2	Consideration for maintain	Piping				
Header type	C	Front-end-branching	Type of piping	Hot-water piping				
Insulated	C	Not insulated	Heat insulation					
90% or more	C	50%	Primary energy efficiency	Water heater				
CASBEE LR$_H$1 3.1: Level 4 or over	C	CASBEE LR$_H$1 3.1: Level 2	Water-saving functions	Water-using equipment				
10% or more of the total water usage	C	0 (Zero)	Rainwater usage	Equipment for rain-water use	Rainwater usage	0 (Zero)	C	10% or more of the total water usage
100% or more	C	60 – 85%	Energy-saving achievement rate	Lighting fixtures & appliances				
Energy usage of the whole home or more	C	0 (Zero)	Harnessed natural energy	Equipment for natural energy	Harnessed natural energy	0 (Zero)	C	Energy usage of the whole home or more
Indigenous species	A	Indigenous species	Species	Garden plants	Fire resistance	High & mid fire resistance	A	High or mid fire resistance
Hedge or Resources-saving materials	A	Hedge & Resources-saving materials	Material	Fence	Form	Not blocking sight line	A	Not blocking sight line
						Not blocking communication	A	Not blocking communication
Places receiving a lot of sunlight	A	Receiving a lot of sunlight	Places in the home	Rooms used at daytime				
				Specified bedroom	Relation with toilet & bath	On the same floor	A	On the same floor
Building close together	A	Building close together	Places in the home	Rooms where water is used				
				Doorways	Difference in level	No differences	A	No differences
					Width	72cm	B	80cm or over
40% or more	A	45%	Ratio to the exterior area	Garden area				

Table 3-2. An example of sustainability checkup on a home as an object

5. Discussion

This study has shown a method for smooth design of practical control systems for sustainable development with a case study. In Chapter 3, we have provided the method, that is, the two-step preparatory work for designing such control systems: (1) determining the relationship between the standard human activities and sustainable development, (2) sustainability checkup on human activities as an object. Chapter 4 has demonstrated a case study, applying this method to homes. This chapter discusses the results of the case study from three viewpoints: (1) the effects of the method on control system design, (2) the value of the case study itself, (3) future work.

5.1 The effects of the method on control system design

The results of the case study have shown that the two-step preparatory work facilitates control system design in three ways, as shown in Fig. 6 (Fujihira & Osuka, 2011; Fujihira, 2011). First, as I intended at the beginning, this method can identify a controlled object, controlled variables, and their desired values. Therefore, it enables system designers to 'identify a controlled object and control objective.'

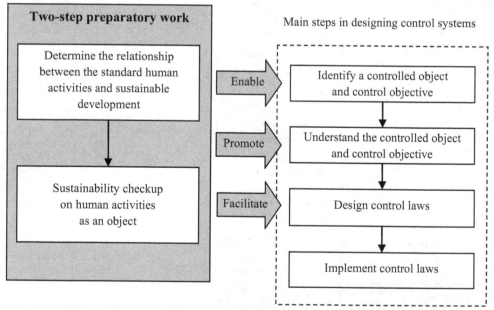

Fig. 6. The effects of the two-step preparatory work on control system design

In addition, this method also promotes 'understanding the controlled object and control objective.' Through sustainability checkup, system designers can comprehensively understand the relationship between important elements of an object and sustainable development. As a result, they can obtain overall and balanced understanding about the controlled object and control objective.

Moreover, this method also facilitates 'designing control laws.' Sustainability checkup enables system designers to understand both what should be controlled and the courses of control so that they can easily design control laws.

5.2 The value of the case study itself

This section examines the value of the case study itself, by comparing it with existing assessment systems for sustainable homes which are used in Japan and the world.

In Japan, *the Japan Housing Performance Indication Standards (JHPIS)* and *Comprehensive Assessment System for Building Environmental Efficiency (CASBEE) for Home (Detached House),* both of which I mentioned in Chapter 4, are used as public performance assessment systems for homes. Japanese Ministry of Land, Infrastructure and Transport provided JHPIS in 2001, aiming to improve housing conditions and sustainability of the built-environment (Building Center of Japan, 2009). JHPIS assesses and indicates housing performance from a variety of angles: structural stability, fire safety, mitigation of degradation, measures for maintenance, thermal environment, indoor air environment, luminous and visual environment, acoustic environment, consideration for the aged and others, security against intrusion (Japanese Ministry of Land, Infrastructure and Transport, 2001). Meanwhile, CASBEE was developed by a committee set up in the Institute for Building Environment and Energy Conservation under the initiative of Japanese Ministry of Land, Infrastructure and Transport. *CASBEE for Home,* one of CASBEE tools, assesses the environmental performance of detached houses from two viewpoints: 'environmental quality (Q)' and 'environmental load (L).' Each of Q and L has three assessment categories: comfortable, healthy and safe indoor environment (Q1), ensuring a long service life (Q2), creating a richer townscape and ecosystem (Q3), conserving energy and water (L1), using resources sparingly and reducing waste (L2), consideration of the global, local and surrounding environment (L3) (Japan GreenBuild Council & Japan Sustainable Building Consortium, 2008).

Other countries of the world are also promoting such assessment systems, including *EcoHomes* of BREEAM in the United Kingdom, *LEED for Homes* in the United States, and *Green Star* in Australia. BREEAM, or Building Research Establishment Environmental Assessment Method, is one of the most comprehensive and widely recognized measures of a building's environmental performance (BREEAM, 2010a). *EcoHomes,* a version of BREEAM for homes, assesses the performance of homes in the following areas: energy, transport, pollution, materials, water, land use and ecology, health and well-being, management (BREEAM, 2010b). LEED, or Leadership in Energy and Environmental Design, is an internationally recognized green building certification system (U.S. Green Building Council, 2011). *LEED for Homes,* a home version of LEED, measures the overall performance of a home in eight categories: innovation and design process, location and linkages, sustainable sites, water efficiency, energy and atmosphere, materials and resources, indoor environmental quality, awareness and education (U.S. Green Building Council, 2008). *Green Star,* which was developed by the Green Building Council of Australia, is a comprehensive, national, voluntary environmental rating system for buildings. Green Star tools, which include Multi Unit Residential, assess nine categories: management, indoor environmental quality, energy, transport, water, materials, land use and ecology, emissions, innovation (Green Building Council of Australia, 2011).

The above public assessment systems contain a variety of essential information; therefore, we referred to them when conducting this case study. On the other hand, as compared with these existing assessment systems, generally, the method in this case study has the following advantages (Fujihira, 2011).

1.   Simplicity and clarity

Table 2 is easy to understand because it simply and clearly shows the relationship between the standard home and sustainable development.

2.   Systematic

Table 2 systematically demonstrates the relationship between the material and spatial elements of the standard home and both the natural environment and humans' well-being. Accordingly, it provides balanced and comprehensive understanding of such relationship.

3.   Ease of use

All of the elements shown in Table 3 are equivalent to real parts of homes. Therefore, when conducting a checkup on a home by using a checkup sheet like Table 3, designers simply check the home's parts which correspond to the elements. As a result, they can easily assess the variables of the elements.

4.   Ease of finding measures for improvement

The results of a sustainability checkup like Table 3 show the elements which should be controlled, controlled variables, and their desired values, all at the same time. Therefore, such checkups enable designers to understand both what should be improved and the courses of improvement and help to find measures for improvement.

The above advantages show that the case study itself has sufficient practical use. Moreover, these advantages indicate the superiority of the method, or the two-step preparatory work for smooth control system design.

5.3 Future work

Our future work includes the following three tasks: (1) further research on sustainable homes, (2) direct support methods for designing control laws, (3) further case studies.

1.   Further research on sustainable homes

Table 2 has successfully demonstrated the essence of sustainable homes, by determining the relationship between important elements of the standard home and sustainability conditions. However, this table probably has a room for improvement. We need to continue making efforts to improve this table through further research on sustainable homes.

In addition, it is also necessary to update this table, as occasion arises. The elements, variables, and their desired values which are shown in Table 2 can be changed or varied, in response to developments in related sciences, innovations in related technologies, and changes in social conditions. Therefore, we need to update this table, responding to such developments, innovations and changes.

2.   Direct support methods for designing control laws

The two-step preparatory work enables system designers to identify and understand a controlled object and control objective as well as helps design control laws. However, our final goal is to establish a methodology of designing control systems for sustainable development. For this purpose, it is also necessary to show methods for supporting the design of control laws more directly.

3.   Further case studies

In this study, we have conducted a case study, selecting the home as a unit of human activities. In order to increase reliability of this method, it is necessary to conduct further case studies. In the future case studies, we should select other units of human activities; for example, the city or town.

6. Conclusion

This study has shown the two-step preparatory work for smooth control system design for sustainable development with a case study. Chapter 3 has provided the two-step method: (1) determining the relationship between the standard human activities and sustainable development, (2) sustainability checkup on human activities as an object. Chapter 4 has applied this method to homes and demonstrated a case study. First, after selecting important elements of the standard home on the basis of the two factors, material and space, we have determined the relationship between such elements and sustainable development. Next, as the second step, we have conducted a sustainability checkup on a home as an object. The results of the case study have demonstrated the effectiveness of this method, for it enables system designers to identify and understand a controlled object and control objective as well as helps them design control laws. Furthermore, the usefulness of the case study itself has also indicated the effectiveness of this method. Our future work includes further research on sustainable homes, showing direct support methods for designing control laws, and further case studies.

7. Acknowledgement

I would like to thank Mr. Vance Carothers for his valuable advice and suggestions on the English expression of this paper.

8. References

Beniger, J.R. (1986). *The Control Revolution - Technological and Economic Origins of the Information Society*, Harvard University Press, ISBN 0-674-16986-7, Cambridge, Massachusetts, USA

BREEAM (2010a). What is BREEAM?, In: *BREEAM*, 03.10.2011, Available from: http://www.breeam.org/page.jsp?id=66

BREEAM (2010b). EcoHomes, In: *BREEAM*, 03.10.2011, Available from: http://www.breeam.org/page.jsp?id=21

Building Center of Japan (2009). Publication in English, In: *The Building Center of Japan*, 03.10.2011, Available from: http://www.bcj.or.jp/en/services/publication.html

Energy Conservation Center, Japan (n.d.). Information on Energy-Saving Products (in Japanese), In: *Energy Conservation Center, Japan,* 03.10.2011, Available from: http://www.seihinjyoho.jp/index.php

Fujihira, K. (1999). *An Introduction to Environmentology,* Nihon-Keizai-syuppansha, ISBN 4-532-14743-3, Tokyo, Japan (in Japanese)

Fujihira, K. (2001). *A Short Introduction to Environmentology,* Kamogawa-syuppan, ISBN 4-87699-626-1, Kyoto, Japan (in Japanese)

Fujihira, K. (2011). An Approach to Sustainable Homes by Applying Control Science, *Advanced Materials Research,* ISSN 1022-6680, online available since 2011/Nov/29, DOI: 10.4028/www.scientific.net/AMR.403-408.2087, Trans Tech Publications, Switzerland

Fujihira, K. & Osuka, K. (2009). An Educational Methodology for Sustainable Development, *ICCAS-SICE 2009 Final Program and Papers,* ISBN 978-4-907764-33-3, Fukuoka, Japan, August 18-21, 2009

Fujihira, K. & Osuka, K. (2010). How do we synthesize control systems for sustainability?, *Proceedings of Society of Environmental Science Annual Convention 2010,* p. 56, Tokyo, Japan, September 16-17, 2010 (in Japanese)

Fujihira, K. & Osuka, K. (2011). An Approach to Designing Control Systems for Sustainable Development, *Proceedings of SICE Annual Conference 2011,* ISBN 978-4-907764-38-8, Tokyo, Japan, September 13-18, 2011

Fujihira, K.; Osuka, K.; Yoshioka, T. & Hayashi, N. (2008). An Educational Methodology for Sustainable Development Applying Control Theory and Confirming its Validity, *Environmental Education,* Vol. 18-1, pp. 17-28, 2008 (in Japanese)

Green Building Council of Australia (2011). What is Green Star?, In: *GBCA,* 03.10.2011, Available from: http://www.gbca.org.au/green-star/green-star-overview/

Japanese Ministry of Land, Infrastructure and Transport (2001). *Japan Housing Performance Indication Standards,* Japanese Ministry of Land, Infrastructure and Transport (in Japanese)

Japan GreenBuild Council & Japan Sustainable Building Consortium (2008). *Comprehensive Assessment System for Building Environmental Efficiency (CASBEE) for Home (Detached House)– Technical Manual 2007 Edition,* Institute for Building Environment and Energy Conservation, Tokyo, Japan

Oxford Dictionaries (2010). *Oxford Dictionary of English Third edition,* Oxford University Press, ISBN 978-0-19-957112-3

U.S. Green Building Council (2008). LEED for Homes Rating System, In: *USGBC,* 03.10.2011, Available from: http://www.usgbc.org/ShowFile.aspx?DocumentID=3638

U.S. Green Building Council (2011). What LEED Is, In: *USGBC,* 03.10.2011, Available from: http://www.usgbc.org/DisplayPage.aspx?CMSPageID=1988

Permissions

The contributors of this book come from diverse backgrounds, making this book a truly international effort. This book will bring forth new frontiers with its revolutionizing research information and detailed analysis of the nascent developments around the world.

We would like to thank Dr. Chaouki Ghenai, for lending his expertise to make the book truly unique. He has played a crucial role in the development of this book. Without his invaluable contribution this book wouldn't have been possible. He has made vital efforts to compile up to date information on the varied aspects of this subject to make this book a valuable addition to the collection of many professionals and students.

This book was conceptualized with the vision of imparting up-to-date information and advanced data in this field. To ensure the same, a matchless editorial board was set up. Every individual on the board went through rigorous rounds of assessment to prove their worth. After which they invested a large part of their time researching and compiling the most relevant data for our readers. Conferences and sessions were held from time to time between the editorial board and the contributing authors to present the data in the most comprehensible form. The editorial team has worked tirelessly to provide valuable and valid information to help people across the globe.

Every chapter published in this book has been scrutinized by our experts. Their significance has been extensively debated. The topics covered herein carry significant findings which will fuel the growth of the discipline. They may even be implemented as practical applications or may be referred to as a beginning point for another development. Chapters in this book were first published by InTech; hereby published with permission under the Creative Commons Attribution License or equivalent.

The editorial board has been involved in producing this book since its inception. They have spent rigorous hours researching and exploring the diverse topics which have resulted in the successful publishing of this book. They have passed on their knowledge of decades through this book. To expedite this challenging task, the publisher supported the team at every step. A small team of assistant editors was also appointed to further simplify the editing procedure and attain best results for the readers.

Our editorial team has been hand-picked from every corner of the world. Their multi-ethnicity adds dynamic inputs to the discussions which result in innovative outcomes. These outcomes are then further discussed with the researchers and contributors who give their valuable feedback and opinion regarding the same. The feedback is then collaborated with the researches and they are edited in a comprehensive manner to aid the understanding of the subject.

Apart from the editorial board, the designing team has also invested a significant amount of their time in understanding the subject and creating the most relevant covers. They scrutinized every image to scout for the most suitable representation of the subject and create an appropriate cover for the book.

The publishing team has been involved in this book since its early stages. They were actively engaged in every process, be it collecting the data, connecting with the contributors or procuring relevant information. The team has been an ardent support to the editorial, designing and production team. Their endless efforts to recruit the best for this project, has resulted in the accomplishment of this book. They are a veteran in the field of academics and their pool of knowledge is as vast as their experience in printing. Their expertise and guidance has proved useful at every step. Their uncompromising quality standards have made this book an exceptional effort. Their encouragement from time to time has been an inspiration for everyone.

The publisher and the editorial board hope that this book will prove to be a valuable piece of knowledge for researchers, students, practitioners and scholars across the globe.

List of Contributors

Bogart Yail Marquez, Ivan Espinoza-Hernandez and Jose Sergio Magdaleno-Palencia
COLEF and ITT, México

Radu Radoslav, Marius Stelian Găman, Tudor Morar, Ştefana Bădescu and Ana-Maria Branea
Faculty of Architecture,"Politechnica" University of Timişoara, Timişoara, Romania

Theresa Glanz, Yunwoo Nam and Zhenghong Tang
University of Nebraska-Lincoln, USA

Svetlana Dj. Mihic and Aleksandar Andrejevic
Faculty of Business and Law Studies, Novi Sad, Serbia

Mirela Mazilu
University of Craiova, Department of Geography, Romania

Beatriz Amarilla and Alfredo Conti
Research Laboratory on the Territory and the Environment (LINTA), Scientific Research Commission, province of Buenos Aires (CIC), Argentina

Lidija Petrić
University of Split, Croatia

Kazutoshi Fujihira
Institute of Environmentology, Japan

Printed in the USA
CPSIA information can be obtained
at www.ICGtesting.com
JSHW011401221024
72173JS00003B/375